THE ECLIPSE
OF COMMUNITY

An Interpretation
of American
Studies

hARpER ⚜ τORChBOO'RS

A reference-list of Harper Torchbooks, classified
by subjects, is printed at the end of this volume.

"Humanism, in every period, must wear a different countenance, since its function at any moment is to turn the course of culture in the direction of men's vital needs. Is it not the task of modern humanism to direct the mind toward the building up of a social group, above the approaching catastrophic anarchy of industrial and national egoisms, but at the same time to defend the human personality against the ever-threatening oppression of the group thus reinforced?"

"And, when we feel inclined to abuse our age, let us beware of excepting ourselves. By consenting to recognize in ourselves the virulent nucleus of the same tendencies that we condemn in our contemporaries we shall be on the way to discovering in them the germ of the protest which has found utterance in us. This will only be an affront to our pride; but at this expense we shall regain some confidence in our species and our period; which is well worth a sacrifice, even the sacrifice of the bitter and arrogant pleasure of being in the right against the world."

—from *The Myth of Modernity*,
   by CHARLES BAUDOIN, pages 78 and 135

THE ECLIPSE
OF COMMUNITY

An Interpretation
of American
Studies

BY MAURICE R. STEIN

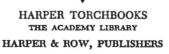

HARPER TORCHBOOKS
THE ACADEMY LIBRARY
HARPER & ROW, PUBLISHERS
NEW YORK

PREFACE TO THE TORCHBOOK EDITION

Since the original edition of this book appeared only four years ago, there is little to be said about it now beyond a few points of clarification. The book, perhaps partly as a result of its title, left certain readers with the impression that my main purpose was to present a pessimistic picture of American community life. This new introduction provides an opportunity to mitigate that impression by allowing me to indicate some of the expansions, contractions and transformations of purpose that occurred during the seven years of work on the book:

1. I undertook the project in the first place and my desire to continue was sustained at every phase because I enjoyed reading community studies far more than most other sociological writing.

2. I found that these studies provided important insight into the social life of earlier generations of Americans as well as into my own experiences and those of my contemporaries.

3. I discovered that I could identify generalizable institutional configurations and generalized patterns of change in the separate studies.

4. I realized that these separate patterns of change could be linked with each other through viewing them as examples of the major social processes studied by classical social theory.

5. I decided to treat these small-scale studies of American communities as case studies showing the ways in which large-scale social processes shaped human affairs in local settings.

6. I took special pleasure in the imaginative descriptions of local settings provided by this literature and wanted to create a theory that would retain some of this descriptive material within a generalized framework.

7. I was struck by the sharply critical attitude which community sociologists sometimes assumed toward the human consequences of the processes they described and wanted to include such critical responses within my theory.

8. I wanted a theory that combined the search for systematic generalizations with a search for significant historical patterns and a search for critical commentary on American community trends.

9. I learned that such a theory had to orient the student of a particular community to the relation between his subject-matter and the totality of community patterns, processes and problems appearing in the same historical period.

10. I found it necessary to invoke a research mood which stressed the study of past forms of American community life in order to help identify emerging forms, emerging problems and potential solutions.

11. I constructed concepts that were sufficiently flexible to permit the sociologist to shape his categories and focus his concerns according to the issues raised by a rapidly changing world.

12. I deepened the bases for interpretation and evaluation by including anthropological and psychoanalytic perspectives which helped illuminate hidden aspects of the relations between men and their communities.

Each reader can decide for himself whether these purposes were satisfactorily accomplished in the book at hand. A detailed statement describing the evolution of this project in more personal terms can be found in my essay in *Reflections on Community Studies* (Wiley, 1964), edited by Joseph Bensman, Arthur Vidich and myself. The entire project was undertaken in the hope that knowledge about the eclipse of community would help sociologists and lay readers alike to better understand, evaluate and ultimately transform the communities in which they live.

Cambridge, Mass.   M.S.
January, 1964

PREFACE AND ACKNOWLEDGMENTS

This book interprets the problems and prospects of American community life, using field studies conducted by sociologists during the past half century. Shapes and forces of change are explored as they fashion the destinies of persons living in settings ranging from small towns to great cities and from impoverished urban slums to prosperous suburbs. Anthropological and psychoanalytic perspectives are used to illuminate deeper meanings of recent trends. The analysis proceeds by comparing the dominant myths in each community with relevant social realities. Familiarizing ourselves with many such patterns helps to nourish the kind of sociological imagination needed for examining the contrasts between myths and realities in our own communities. We should not let the fact that we cannot predict the exact outcome of the present eclipse dissuade us from trying to see as far ahead as we possibly can.

I would like to take this opportunity to express gratitude to my teachers at the University of Buffalo, especially Alvin Gouldner and the late Nathaniel Cantor; and to Robert Merton and Seymour Fiddle of Columbia University. Other people who contributed to the progress of this book, sometimes without knowing they were doing so, include Conrad Arensberg, Jacques Barzun, George Eaton Simpson, Daniel Stein, Ruth and Robert Paradise, Lawrence and Myrtle Wilson, Irene Pierce, Thalia Howe, Ruth Tendler, Gerhard and Hanna Hoffman, and Franklin Carter.

Brandeis University provided an exceptionally congenial atmosphere in which the bulk of the writing was completed. Irving Howe gave sympathetic editorial advice of an indispensable kind; Stanley Diamond and the late Paul Radin broadened my conception of social science in vital respects; and Max Lerner offered generous encouragement and assistance. Arthur Vidich and Joseph Bensman confirmed several of my hypotheses about American communities and commented most helpfully on the manuscript. Suzanne Keller and Robert Feldmesser were kind enough to read an early draft

CONTENTS

CONTENTS

INTRODUCTION

AMERICAN HISTORIANS who have been writing the history of the past fifty years have access to an important source of data in the many community studies completed by American sociologists during this period. These studies differ from other sources of information about national history in that they describe events occurring at a great distance from the places where national power was concentrated. No picture of the daily round of life in a middle-sized American town during the twenties can be found to improve upon that presented by the Lynds in *Middletown,* while any historian interested in the social effects of the depression can learn much from *Middletown in Transition.* Community studies cannot provide information about the men and events on the national scene that influence historical processes so decisively, but they do describe the effects of these processes on the everyday lives of ordinary men and women. They fill out the historical record by giving the intimate meanings that large-scale changes had for a limited segment of the population.

In doing this, it is clear that they have serious limitations. Thus, a historian would not be likely to assume that the effects of the depression in Muncie, Indiana, were comparable to its effects in Detroit, Michigan. The usefulness of a community study as a source of information about national experiences and responses is limited by the extent of its representativeness. These studies actually fall into a kind of limbo between sociology and history. They are usually drawn on for illustrative purposes but rarely incorporated into the narrative interpretations of historians, on the one hand, or into the general theories of sociologists, on the other. Before either sociologists or historians can use these studies to the fullest possible extent, some kind of general framework relating the studies to each other has to be devised. This book will present one such framework.

In formal terms, the purpose of such a framework is to raise low-order generalizations about specific events and processes occurring in one community during a circumscribed time period to a higher level, so that they can be related to other low-order generalizations about similar events occurring in a different community during the same or another time period. In less formal terms, the framework aims at developing an approach to community studies which will discern similar processes at work in different contexts. Or, to put it still another way, the purpose is to devise an approach to community studies in which each investigation becomes a case study illustrating the workings of generalized processes in a specific setting.

The range of variation found among the many community studies in the literature is great, making the task of reconciling them difficult. Aside from obvious differences between the communities in terms of such important dimensions as age, size, economy, regional location, etc., additional problems are introduced by the fact that the field work was conducted at different times and that the talents, training, facilities, and interests of the field workers differed widely from study to study. Each research report is a synthesis by the author of several orders of data about a particular community, arranged according to his sense of significant social structures and processes. This synthesis rarely takes into account the relation of the material to other related studies nor do most research reports contain chapters which satisfactorily present generalized conclusions. The community sociologist has been a better ethnographer than a theorist and this is probably as it should be. Weaving the scattered strands of a single community into a coherent picture is in itself a difficult task.

Achieving the formal purpose of this book—that is, the development of a framework for relating disparate community studies to each other—presupposes the substantive purpose, which is to develop a theory of American community life. The purposes are closely linked since the perception of relationships between studies rests upon the assumption that similar social forces are at work in the separate communities. Social theories about the forces transforming Western society during the past four centuries converge on three kinds of processes— urbanization, industrialization, and bureaucratization, as the central sources of change. Such theorists as Marx, Weber, Durkheim, Simmel, and Mannheim may emphasize one or the other, but all recognize the three as being involved in fundamental social change. On the basis of this agreement, and the conviction that all three could be observed at work in American communities, the decision was made to search for their effects as these are reported in the various community studies examined.

The book is divided into three parts. Part I contains the main portions of the theory of community development based on careful scrutiny of changes caused by urbanization in Chicago, industrialization in Muncie, and bureaucratization in Newburyport during the first thirty-five years of this century. It uses the studies of Robert Park, Helen and Robert Lynd, and Lloyd Warner as the source of patterns in each instance. Part II contains five chapters, each of which applies the case method and substantive hypotheses presented in Part I to studies of different types of communities. Studies of the slum, Bohemia, Southern towns, military communities, and suburbia are covered. Part III returns to general issues and traces the way in which urbanization, industrialization, and bureaucratization shape contemporary American community life.

The plan of study adopted involves distinguishing between generalized social processes and their concrete embodiments in specific communities. For example, Louis Wirth's definition of urbanization as the coming together of a large heterogeneous population within a relatively small area is placed in the context of the studies of Chicago neighborhoods undertaken by Park's students during the twenties and thirties. These studies explored the kinds of natural areas, social disorganization, social control, and social reorganization that followed the rapid influx of immigrant and rural population elements into Chicago between 1900 and 1930. Wirth's conception refers to an abstract set of social processes with highly generalized social consequences, while the empirical studies of Chicago patterns show how the particular pattern of urbanization appearing in America during the first few decades of this century reshaped a particular city. Similarly, if industrialization is defined in terms of the twin processes of mass production and mass consumption, then we can study its occurrence anywhere in the world. But the Lynds allow us to examine the social dislocations occasioned by the specific form of industrialization that transformed American community life around the turn of the century. The reader must be careful, then, to distinguish the generalized process from the concrete embodiment in a specific community. The two are related but distinct.

From the start, it was assumed that the theoretical structure would be a loosely woven one. The very nature of the materials that were being worked with made this inevitable. All of the studies included, except those undertaken as part of a larger scientific enterprise such as the research done by Park's students or the two Middletown volumes, were completed quite independently of each other. This does not mean that sociologists entering a community in the forties, for example, did not read the classic studies of the twenties and thirties. We can assume that they did. But there is always a large inductive element in any community study. The field worker establishes his picture of the social structure gradually as he accumulates relevant details. Only later is he likely to perceive relationships to earlier studies, even though his familiarity with them may have made him sensitive to structural patterns of considerable importance in the first place. William Foote Whyte, in the appendix to the second edition of *Street Corner Society*, offers an illuminating glimpse of the working processes of the sociologist in the field. He shows how seemingly disparate observations congeal into conceptual interpretations when skillfully processed by a trained mind.

[6]

Community sociologists have a ready-made general orientation to their field materials in the functional conception of social institutions. The Lynds, taking their lead from the anthropologist W.H.R. Rivers, divide the first Middletown volume into six major areas of life (i.e., getting a living, making a home, training the young, using leisure, engaging in religious activities, and engaging in community practices). As a rough ethnological guide to the categories of material to be covered in any community, this serves a useful purpose. It is worth noting, however, that more complicated categorical systems must be developed to cover aspects of community life not explicitly included in the scheme. And these added patterns are often significant because they include the distinctive structural features of the community being studied. So, in the second volume, *Middletown in Transition*, which covers the depression period, the Lynds added a long chapter on "caring for the unable" which was dealt with summarily under the more general heading of "engaging in community activities" in the earlier volume. The elevation of this new community institution in the form of federal, state, local, and private "relief" programs symbolized important changes wrought by the depression. Ignoring the conventional categories to focus only on novel patterns also holds dangers. William Foote Whyte emphasizes the distinctive slum social system consisting of patterned relations between the police, politicians, racketeers, and corner boy gangs in *Street Corner Society*, but neglects the family and religion. In his appendix to the second edition, he expresses regret that he failed to include these latter subjects and offers reasons for considering this institutional classification scheme to be an irreducible minimum ethnological standard.

The theory of communities presented in this book has two purposes. First, it aims at making the significant findings of community studies accessible by interpreting these findings within the larger context of changing American society. Secondly, it aims at analyzing present-day community trends to show their relation to earlier trends, to contemporary social problems, and to the quality of modern life. Throughout the book, *theorizing* refers to the process of analyzing community trends. This process has two products, the *theoretical framework*, which includes generalizations about procedures for conducting such analyses, and the *substantive theory*, which includes both generalizations about trends in particular communities and generalizations about American communities as a whole. Readers who are interested in the methodological assumptions underlying the book may want to read the Epilogue first before starting Part I.

On first reading a book like *Middletown*, or any good community study, the ordinary reader is likely to be overwhelmed by the mass of detailed facts included. Observations pile upon observations so that the guiding threads tend to get lost. Very often the unifying themes are not introduced until the final chapters, if they are made explicit at all. This forces the reader to retrace his steps through the book to see the larger unit perceived by the author, but it has the advantage of allowing him to form his own independent impressions. Community studies cannot be read like geometry textbooks in which the argument proceeds from postulates to inferences in exact sequence. Instead, the reader has to allow his impressions of the social structure to grow gradually—*quite as does the field worker in the original situation*. Details have to be mulled over as their meaning changes with shifts in context, and general impressions must be tested by renewed inspection of the reported data.

To repeat a point made earlier, each work has to be viewed as a case study. This means that the "facts of the case" have to be confronted fully before they can be realigned to reveal relationships not emphasized by the author. Since the theory being developed in this book deals with historical processes, applying it to particular studies will always involve bringing the changes in the communities under consideration into focus.

[8]

Either Middletown volume could have been called *Middletown in Transition* and the significance of the second volume lies exactly in the fact that it raises the further question: "What were the effects of the depression on the set of changes observed in the earlier study?" Proper analysis of such transitions in terms of this theory rests upon careful elucidation of the changes and changing sub-groups found by the field worker. The theory can never provide the substantive details; at best, it does help to place them within a broader interpretive context.

This book should facilitate systematic investigation of community processes in historical perspective. To do that, it uses the classical community studies as specimens illustrating the workings of these processes within a special context. By emphasizing the context, it focuses attention on the variety of outcomes which the same processes can achieve without neglecting the equally significant similarities. The unit of study is always the total community. Park liked to think of Chicago as a "social laboratory" for social scientists. This study makes an effort at identifying the limiting conditions surrounding the use of any community for such purposes. In doing so, it aims at illuminating patterns of development in the past and stimulating research on present developments. The point is to show continuity in community patterns and community studies. With this end in mind, the book could have been subtitled, "How to Read a Community Study." But the qualification must be added that from the vantage point of this book the difference between reading a community study properly and doing one well is not nearly as great as has often been supposed.

PART I. FOUNDATIONS

CHAPTER 1

ROBERT PARK AND URBANIZATION
IN CHICAGO

THERE is no more comprehensive approach to the study of American communities than that developed at the University of Chicago during the twenties. Here a group of scholars completed a set of empirical studies that leave us with more sociological knowledge about the city of Chicago than is currently available for any other single American community. In order to appreciate the value of this information, it is important to consider the historical circumstances under which it was gathered. While the Chicago sociologists tended to assume that their findings would have unlimited applicability, we now know that the process of urbanization they observed had certain special features which must be identified before its generalized relevance can be assessed. Certainly their descriptions of cultural areas are no longer satisfactory. Contemporary American ghettos, for example, now differ considerably from those studied so perceptively by Louis Wirth. The central problem facing us in this chapter is the development of a theory of urbanization broad enough to include these studies and also to take account of later material and developments.

Those of us not fortunate enough to have been at the University of Chicago during the twenties can still easily see how this must have been a tremendously exciting atmosphere for social scientists. On the scene during this period were men like William I. Thomas, George Herbert Mead, and Robert Park, ably assisted by Ellsworth Faris, Ernest Burgess, Roderick McKenzie, William Ogburn, Harold Lasswell, Charles Merriam, and T.V. Smith. Students, assistants, and occasional faculty present included Louis Wirth, Herbert Blumer, Everett Hughes, Leonard Cottrell, Franklin Frazier, Emory Bogardus, Herbert Gosnell, Samuel Stouffer, Francis Merrill, Pauline Young, Ruth Cavan, Clifford Shaw, Frederick Thrasher, Harvey Zorbaugh, Paul Cressey, Norman Hayner, and Robert Redfield—to name only those that come immediately to mind.

It is hardly necessary to include here a detailed statement of the still-continuing contribution of these men to sociology. Their interest in urbanization led them to undertake studies that gave shape and substance to most of the current subdivisions of contemporary American sociology. Such areas as urban sociology, the family, criminology, race relations, social problems, social change, sociology of the mass media, public opinion, sociology of occupations, political sociology, social psychology, and social psychiatry all received considerable impetus, if not initial definition, from their theoretical and empirical efforts. Nineteen of the thirty-five presidents of the American Sociological Society received their doctorates from Chicago, most of them during the twenties and early thirties.

Some sense of the temper of the Chicago department can be gained by examining any of the volumes in the University of Chicago sociology series, each with its stock of cross references to other published or soon-to-be published books in the series, and each with its preface by Park or Burgess or Faris. Here was an atmosphere of true continuity and accumulation, with the key to both largely in Park's hands. Research was organized so that even the lowly graduate student could conduct field studies that made immediately significant contributions to the rapidly advancing frontiers of social science. Publication facilities, whether through the University of Chicago Press or the Chicago-controlled *American Journal of Sociology*, were readily available. The city, with its constantly changing surface, was there waiting to be explored. The Chicago group cherished a vision of a developed science of the community which could chart patterns of change so that men might finally fashion their social environments to conform more closely with their ideals.

Many of these sociologists, like the large majority of the population of the time, had spent their early years in small towns. This was the form of community living with which they were most familiar and it provided a frame of reference within which the highly dissimilar features of Chicago social life were perceived and evaluated. It is hard for us today to recapture those experientially rooted perceptions—not only because there are few American cities growing as rapidly as Chicago did in the early twenties, but also because several decades of urban or urbanized-rural living blind us to whatever diversity remains in the cities where we make our homes. This mental block against responding to multitudinous stimuli which Simmel and Park deemed characteristic of the urban personality type has become part of the "social character" of sociologists acclimated to urban life and undoubtedly serves to blind us to its real diversity. The pressure to maintain a privacy that might preserve social distances between people who have no established social ties is deeply internalized and stifles much of the curiosity about strange styles of life that led the farm-reared Chicago sociologists to prowl the streets of their city in search of this very strangeness.

Park himself came from and spent his early years in a small town. Only later did he abandon it for the world of cities, whose scientific interpreter he ultimately became. He studied under John Dewey at Michigan, William James at Harvard, and Georg Simmel at Berlin. His travels took him to many countries, first as a newspaper reporter and later as a distinguished teacher of sociology. Perhaps the most colorful and perceptive report on Chicago, when it was undergoing rapid urbanization, is contained in Park's seminal essay *The City—Suggestions for the Investigation of Human Behavior in the Urban Environ-*

[15]

ment, which first appeared in the *American Journal of Sociology* in 1915 and has since been reproduced in nearly every collection of readings on urban sociology. This essay outlines many of the research projects later carried out with brilliant results by his students, and the best testimonial to its scientific excellence as distinct from its literary value lies in the fact that it still contains a great many unexplored leads for further research and study, a surprising number of which appear to deserve serious attention.

Before we delve further into the orientation of Park and his colleagues, a few statistics may suggest something about the nature of their social experiences. According to the 1950 *World Almanac*, the population of Chicago grew as follows: 1860, 112,172; 1900, 1,698,575; 1910, 2,185,283; 1920, 2,701,705; 1930, 3,376,438. Cold figures alone cannot suggest what it must mean to people living in a city when the population swells at a rate of about one-half million per decade for three consecutive decades. Small wonder that the Chicago sociologists focused on the absence of established institutional patterns in so many regions of the city, stressing that the neighborhoods grew and changed hands so rapidly that sometimes the only constant feature appeared to be mobility. If to this is added the fact that much of the incoming population consisted of foreigners arriving in the heavy wave of immigration to America during the first quarter of this century, while most of the rest were farmers as unaccustomed to city ways as their foreign-born neighbors, it then becomes clear why "disorganization" accompanied "mobility."

While it is occasionally true, as has been charged, that the rural backgrounds of the Chicago sociologists did lead them to view their sprawling city with a slightly jaundiced eye, on the whole these men saw the rewards as well as the costs of urban living. Being "marginal" they were too close to the city as well as to the country to idealize either. The following passage from Park captures something of the subtlety of his insight and seems representative of the views he passed on to his students: "In the freedom of the city every individual, no matter how eccentric, finds somewhere an environment in which he can expand and bring what is peculiar in his nature to some sort of expression. A smaller community sometimes tolerates eccentricity, but the city often rewards it. Certainly one of the attractions of a city is that somewhere every type of individual—the criminal and beggar, as well as the man of genius—may find congenial company and the vice or the talent which was suppressed in the more intimate circle of the family or in the narrow limits of a small community, discovers here a moral climate in which it flourishes."[1] This passage contains many of the essential themes in Park's conception of urban life. It foreshadows the central problem of social control arising out of the necessity for regulating behavior in these diverse "moral regions" without wiping out their individuality. This "freedom of the city" is both immensely attractive and terribly menacing. The city provides supportive subcommunities for the criminally eccentric, as well as for persons whose eccentricities take more socially desirable forms.

One of the central problems of Park's urban sociology is that of identifying control mechanisms through which a community composed of several quite different subcommunities can arrange its affairs so that each of them maintains its own distinctive way of life without endangering the life of the whole. Park wanted to preserve unity in the context of diversity instead of the custom-bound homogeneity of rural life. He was aware of the difficulties involved but clearly felt that the advantages in terms of human freedom and fulfillment outweighed the disadvantages.

Durkheim dealt with the same problem in his interpretation of the transition from mechanical solidarity to organic solidarity as the basis for social order in society. Possible only in a rela-

[1] Park, Robert E., *Human Communities*, The Free Press, Glencoe, Illinois, 1952, p. 86.

[17]

tively undifferentiated and unchanging society, mechanical solidarity depended upon the universal internalization of a common moral code in order to regulate all social interaction. As social differentiation proceeds, a single code is no longer satisfactory for shaping the behavior of the emerging sub-groups. Like Park, Durkheim saw each sub-group developing its own moral code. If the larger social framework was to be preserved, Durkheim felt that these special moral codes must be such as to define the responsibilities of each sub-group to the whole society and thereby articulate their activities with that of those performed by the others. The main social problem becomes finding mechanisms whereby this organic solidarity can be achieved.

There is no question but that Park was seriously influenced by Durkheim. Both were aware of the precariousness of organic solidarity. Durkheim's analysis of suicide, in which he shows how social differentiation releases men from group ties and moral restraints only to leave them isolated and without values that confer meaning on life, underlies much of the theorizing of the urban sociologists. Park too saw that freedom from group constraints could also entail freedom from group supports; his analysis of social relations in the rooming-house district takes up this point. Both were aware of the possibilities for personal disorganization arising from rapid change in standards and opportunities. Durkheim called this "anomie," while Park dealt with it under such headings as "individualization" and "mobility." Personal disorganization was seen by both as the result of exposure to conflicting standards; insofar as cities provided much opportunity for living "on the margin" between different sub-cultures, the likelihood of this form of disorganization was maximized.

Park was always conscious of the creative as well as the disorganizing potentialities of marginality. He saw that marginal people could react to their condition with increased insight into the various sub-cultures to which they were exposed, so that sometimes creative cultural innovation might be achieved as a direct result of this exposure. Human potentialities could be released that would find no place in more homogeneous cultures. Some of these potentialities might prove destructive, while others could constitute genuine advances in patterns of human living. The price for maintaining a society that encourages cultural differentiation and experimentation is unquestionably the acceptance of a certain amount of disorganization on both the individual and social level. Disorganization is desirable only insofar as it serves as a stimulus for creative reorganization, thereby leading to the discovery of more complicated life styles with greater possibilities for human satisfaction. There is always the danger that widespread disorganization will threaten the main values and even the continued existence of the society. The "organic" society and the urban community must be constantly alert to the need for keeping disorganization under control. without either eliminating it or allowing it to eliminate the bases for social order.

Where Durkheim dealt with whole societies, Park was interested in communities. While Park occasionally used the concept to refer to a generalized aspect of social organization, it is clear that his main referent for his conception of community was the city of Chicago. His approach to urban communities never quite managed to take into account the fact that any given nation would have within its boundaries many separate communities in various stages of urbanization. Had he done so, he would have been more concerned with comparing Chicago to other American cities in order to see the differences as well as the similarities in their adaptations to the population influxes caused by their exposure to urbanization. This concern implies a corollary concern with *patterns of urbanization* and opens up the whole set of problems occasioned by obvious differences in the growth process of cities in various national contexts during different historical periods. The relationship between an entire society and the communities of which it is

[19]

composed becomes an object of study and should not be concealed within a general term like "urban society," which ends discussion at a point where it could well begin.

Durkheim's interpretation of the mechanisms of social organization in a complex society are still extremely useful. He assumed that occupational groups would develop their own codes of ethics through which the behavior of their members would be restricted in the interests of the whole society. In this way, the interests of the occupational sub-group and the necessities of "organic solidarity" could be reconciled. Approaching the same problem from the point of view of the community, Park identified a number of other kinds of subgroups whose behavior had to be regulated as well as agencies for accomplishing this regulation. His starting point was existing social disorganization in the city of Chicago and his main structural units the various subcommunities in which this disorganization appeared as well as the agencies of "secondary control" which tried to keep it from getting out of hand.

There is no intention of presenting this interpretation of Park's theorizing as if it were the *only* possible interpretation. I am all too aware that it takes into account some emphases in his work while neglecting others. The purpose of this interpretation is to reformulate the sociological aspects of his theory of urbanization in such a fashion that it can be used to study contemporary communities. Doing so necessarily involves returning to the historical situation out of which the theory was developed, in the hope that this will permit disentangling those aspects of his conception of urban social structure and processes which depend upon conditions in the city of Chicago during the twenties from those which accompany the process wherever it occurs. However, the search for generality will not be conducted according to the model employed by Louis Wirth in his famous essay *Urbanism As a Way of Life*, which seeks to: ". . . discover the forms of social action and organization that typically emerge in relatively permanent, compact settlements of large numbers of heterogeneous individuals."[2] Nor will the plan of the ecologists who search for typical patterns of spatial arrangement be adopted. A less generalized approach, having as its main point of departure the problems of disorganization and social control in Chicago of the twenties, is our choice. The reasons for this should become clear as the argument unfolds; however, it may be pertinent to note that the strategy rests upon the assumption that the most satisfactory way of developing sociological generalizations is through specification of historical contexts. The sociological fruits of Wirth's "general" theory of urbanism or of the highly abstract theorizing of the ecologists, though certainly not inconsiderable, seem less satisfactory when confronted with the problems of social control and social change facing modern communities.

First, then, we must turn to Park's conception of the "natural areas" of the city. These are the various sub-communities that he noted in Chicago and other cities with which he was familiar that served as vehicles for urban freedom as well as for urban disorganization. It is this thread of Park's theory, taken up by many of his students, which provided the best sociological data of all the research undertaken by the Chicago school. The rest of this chapter is devoted to collecting some of the generalizations advanced by these students and showing their relation to our version of Park's theory of urbanization. Beginning, then, with his conception of the city:

"The urban community turns out, upon closer scrutiny, to be a mosaic of minor communities, many of them strikingly different one from another, but all more or less typical. Every city has its central business district; the focal point of the whole urban complex. Every city, every great city, has its more or less exclusive residential areas or suburbs; its areas of light and of heavy industry, satellite cities, and casual labor mart, where men are recruited for rough work on distant frontiers, in the

[2] Wirth, Louis, in *Reader in Urban Sociology*, Hatt, P.K., and Reiss, A. Jr., editors, The Free Press, Glencoe, Illinois, 1951, p. 37.

[21]

mines and in the forests, in the building of railways or in the borings and excavations for the vast structures of our modern cities. Every American city has its slums; its ghettos; its immigrant colonies, regions which maintain more or less alien and exotic culture. Nearly every large city has its bohemias and hobohemias, where life is freer, more adventurous and lonely than it is elsewhere. These are the so-called natural areas of the city."[3]

Here it is clear that "natural areas" were the key to Park's view of the city. They are the building blocks, the inevitable sub-communities, from which the urban constellation arises.

Exactly how "inevitable" they really are is never closely examined. In the above passage, Park notes that "nearly every large city has its bohemias and hobohemias," leaving open the questions of whether smaller cities have them too or the reasons why a "few" large cities do not have them. However the assumption of inevitability is not without underpinning: "They [natural areas] are the products of forces that are constantly at work to effect an orderly distribution of populations and functions within the urban complex. They are 'natural' because they are not planned, and because the order that they display is not the result of design, but rather a manifestation of tendencies inherent in the urban situation; tendencies that city plans seek—though not always successfully—to control and correct. In short, the structure of the city, as we find it, is clearly just as much the product of the struggle and efforts of its people to live and work together collectively as are its local customs, traditions, social ritual, laws, public opinion, and the prevailing moral order."[4]

The "natural areas" recur because they allow city people to satisfy fundamental needs and solve fundamental problems. So long as the population elements with the relevant needs and problems remain, the areas will probably be found. Further-

[3] Park, op.cit., p. 196.
[4] Ibid.

more, once established, these sub-communities have a tendency to perpetuate themselves: "In short, the place, the people, and the conditions under which they live are here conceived as a complex, the elements of which are more or less completely bound together, albeit in ways which as yet are not clearly defined. It is assumed, in short, partly as a result of selection and segregation, and partly in view of the contagious character of cultural patterns, that people living in natural areas of the same general type and subject to the same social conditions will display, on the whole, the same characteristics."[5] "Natural areas," then, not only select residents whose needs and problems are congruent with local institutional facilities; they also socialize their inhabitants, both young and old, along appropriate lines.

Sociologists then expect to find typical institutional patterns and sub-cultural value systems bound up with personality types into a functioning whole that constitutes the sub-community or "natural area": "Every natural area has, or tends to have, its own peculiar traditions, customs, conventions, standards of decency and propriety, and, if not a language of its own, at least a universe of discourse, in which words and acts have a meaning which is appreciably different for each local community. It is not difficult to recognize this fact in the case of immigrant communities which still preserve more or less intact the folkways of their home countries. It is not so easy to recognize that this is true in those cosmopolitan regions of the city where a miscellaneous and transient population mingles in a relatively unrestrained promiscuity. But in these cases the very freedom and the absence of convention is itself, if not a convention, at least an open secret."[6] Thus, Park assumes that the various urban sub-communities will establish social conventions to regulate the behavior of their members. He recognizes the important possibility that regions appear in which isolation

[5] *Ibid.*
[6] *Ibid.*, p. 201.

[23]

and the absence of shared ties become the distinguishing feature of the area.

However, this segregation into "natural areas" cannot be treated as a static phenomenon. There is a constant sorting process in progress which extends even to the sub-communities themselves as well as to the people in them. Within the immigrant ghettos, for example, there is a tendency for the more successful members to move to better (i.e., newer, wealthier) neighborhoods. This movement usually took place as part of a collective migration so that areas of "second settlement" appeared with their own diluted version of the "first settlement" social and cultural patterns:

"Such segregations of population as these take place, first, upon the basis of language and of culture, and second, upon the basis of race. Within these immigrant colonies and racial ghettos, however, other processes of selection inevitably take place which bring about segregation based upon vocational interests, upon intelligence, and personal ambition. The result is that the keener, the more energetic, and the more ambitious very soon emerge from their ghettos and immigrant colonies and move into an area of second immigrant settlement, or perhaps into a cosmopolitan area in which the members of several immigrant and racial groups meet and live side by side. More and more, as the ties of race, of language, and of culture are weakened, successful individuals move out and eventually find their places in business and in the professions, among the older population group which has ceased to be identified with any language or racial group. The point is that change of occupation, personal success or failure—changes of economic and social status, in short—tend to be registered in changes of location. The physical or ecological organization of the community, in the long run, responds to and reflects the occupational and the cultural."[7]

This passage is especially important because it raises the

[7] *Ibid.*, p. 170.

possibility that some kind of developmental models encompassing changes over time are required to understand the urban constellation of "natural areas." It is the opening wedge for more detailed historical interpretations.

Park goes on to suggest that the theory of "natural areas" be used to analyze distinctive social characteristics distributed throughout the city. In particular it provides a framework for organizing studies of types of deviant behavior as these cluster in various sub-communities:

"The difference in sex and age groups, perhaps the most significant indexes of social life, are strikingly divergent for different natural areas. There are regions in the city in which there are almost no children, areas occupied by the residential hotels, for example. There are regions where the number of children is relatively very high: in the slums, in the middle-class residential suburbs, to which the newly married usually graduate from their first honeymoon apartments in the city. There are other areas occupied almost wholly by young unmarried people, boy and girl bachelors. There are regions where people almost never vote, except at national elections; regions where the divorce rate is higher than it is for any state in the Union, and other regions in the same city where there are almost no divorces. There are areas infested by boy gangs and the athletic and political clubs into which the members of these gangs or the gangs themselves frequently graduate. There are regions in which the suicide rate is excessive; regions in which there is, as recorded by statistics, an excessive amount of juvenile delinquency, and other regions in which there is almost none."[8]

Social statistics collected so as to highlight age and sex concentration patterns in urban sub-communities allows the analyst to detect important relationships that remain concealed when rates for the whole city are reported. Interpreting these relationships rests upon showing how the statistical fact grows

[8] *Ibid.*, p. 172.

out of the institutional structure of the particular sub-community so that excessive delinquency, for example, would have to be explained in terms of institutional patterns like juvenile gangs, poverty, broken homes, and the like.

Under the conditions of urbanization observed in Chicago during the twenties, some of the sub-communities seemed to be especially lacking in effective social controls. All of them appeared to be undergoing some kind of general loosening in their moral codes, but this had gone farthest in slum areas. The spread of disorganization was the direct result of the mobility and individualization accompanying city growth:

"But with the growth of great cities, with the vast division of labor which has come in with machine industry, and with movement and change that have come about with the multiplication of the means of transportation and community, the old forms of social control represented by the family, the neighborhood, and the local community have been undermined and their influence greatly diminished.

"This process by which the authority and influence of an earlier culture and system of social control is undermined and eventually destroyed is described by Thomas—looking at it from the side of the individual—as a process of 'individualization.' But looking at it from the point of view of society and the community it is social disorganization.

"We are living in such a period of individualization and social disorganization. Everything is in a state of agitation—everything seems to be undergoing a change. Society is, apparently, not much more than a congeries and constellation of social atoms."[9]

But even in this period of widespread disorganization, Park saw new institutional forms emerging to maintain social order. These include such "secondary" agencies as the police and law courts, the newspapers, schools, and settlement houses, which spread "proper" values among potential offenders; and the stock

9 *Ibid.*, p. 59.

market or the political machine, which adjudicate between various interest groups. Furthermore, some of the behavior classified as "disorganized," such as the rising divorce rate, is susceptible to reinterpretation as a distinctive new form of "organization," insofar as divorce can be followed by remarriage based on more rational criteria of choice.

To summarize the place of the theory of "natural areas" in Park's work, his comments about its methodological function are especially revealing: "The natural areas of the city, it appears from what has been said, may be made to serve an important methodological function. They constitute, taken together, what Hobson has described as 'a frame of reference,' a conceptual order within which statistical facts gain a new and more general significance. They not only tell us what the facts are in regard to conditions in any given region, but insofar as they characterize an area that is natural and typical, they establish a working hypothesis in regard to other areas of the same kind."[10]

It is clear, from this essay at least, that Park hoped to build a set of hypotheses and generalizations about each of the various "natural areas" of Chicago which could be applied to other cities as well:

"The possibility of drawing inferences from what has been observed and described in London, as to what we might expect in New York or Chicago, it should be said, rests on the assumption that the same forces create everywhere essentially the same conditions. In practice it might turn out that this expectation would not be, or would seem not to be, justified by the facts. It could at least be verified, and that is the main point. Should it turn out that the expectations in regard to London, based upon studies in New York and Chicago, were not justified by the facts, this would at least raise the question as to how far the forces that made London what it is were different from those that made Chicago and New York. And this would lead,

[10] *Ibid.*, p. 198.

[27]

in turn, to a more thoroughgoing and a more accurate analysis of the actual forces at work in both instances.

"Thus the result of every new specific enquiry should re-affirm or redefine, qualify or extend, the hypothesis upon which the original enquiry was based."[11]

Sociological literature hardly contains a more sophisticated statement of the desired relation between theory and research in scientific inquiry. For us, it provides a valuable criterion for assessing the cumulative contributions made by Park's students in their studies of specific sub-communities and disorganization patterns.

We have found it necessary to quote at length from Park's essays because the development of urban sociology has not followed the lines indicated in them. "Natural areas" as an object of inquiry have received little attention in recent years and the whole social-anthropological field tradition, which figures so large in Park's methodology, has given way to statistical and ecological research. Everett Hughes and his students remain the most important descendents of Park now working in the field of sociology. They take occupations as their point of departure, but consistently keep in mind the urban setting within which the various occupational groups and career patterns take their form.

One reason for the lack of concern with "natural areas" stems from the fact that the early area studies seem to have little bearing on modern urban life. Since, thirty years later, the ghettos and the slums are quite different places from those studied by Wirth, Zorbaugh, and Shaw, it is not easy to see that these studies contribute to a developing theory of urban social organization which is as relevant to our time as to theirs. This difficulty stems from the failure to work out carefully the relations between special conditions existing in Chicago during the first thirty years of this century and other possible patterns of urbanization during that time and thereafter. Grasping the

11 *Ibid.*

generalized applicability of this theory depends most of all on understanding the specific context from which it was evolved.

In other words, the whole set of studies of Chicago and its sub-communities conducted by Park's students can be viewed as contributions to a single case study of the effects that urbanization in a particular form had on a particular community during a specific interval of time. Park assumes that other communities undergoing this same process will experience similar problems and develop similar structures that can be regarded as "natural areas." It is unfortunate that neither he nor his students saw fit to conduct systematic research in different cities of a sociological rather than an ecological nature. Other community studies such as the Lynds' *Middletown* show that the effects of urbanization during the first quarter of the century on isolated smaller communities was less severe and different from that felt by Chicago.

Once this is seen, it becomes obvious that the urbanization process must always be defined in specific substantive terms. Park's generalized propositions referring to changes in social organization that emerge with rapid population influx must be carefully qualified according to the composition of the immigrant population, its size in relation to the size of the community entered, and the period of time over which the entry is accomplished. All of these are subject to independent variation as is the national context in which urbanization occurs.

Where, then, does this leave Park's theory of urbanization and especially his hypotheses about "natural areas"? They remain extremely useful as guides to the general problems of community organization and disorganization in American cities during the twenties. Insofar as this is the case, they constitute the setting against which contemporary problems of community organization arise. In studying a modern community, it is necessary to understand the kinds of population influxes that occurred over the last fifty years. The history of contemporary sub-communities must be grasped in order to elucidate patterns of change, as well as to understand the past experiences that condition the present-day behavior of their members. Such a modified theory of urbanization sensitizes the sociologist to an important process affecting community life as it has developed historically and provides a basis for analyzing the effects of this process at different points in time and space.

To see more clearly how Park's focus on "natural areas" is related to his theory of urbanization, let us look at the slum, the area studied most closely by the Chicago sociologists. The main reason for choosing the slum was probably because it sheltered the most salient kinds of disorganized behavior to be found in Chicago. Slums embodied to the fullest extent novel institutional patterns that, to a lesser degree, were also emerging in the more stable urban sub-communities. Here were found the greatest loss of control by parents over children, as well as the greatest freedom from middle-class social norms. Since the slums were closely linked to ethnic ghettos, they became the stage for the main dramas of social life characterizing this period of American community development. Furthermore, insofar as urban sociologists wished to develop practical theories to be used by men of affairs concerned with improving community life, the slum certainly was the sub-community that most clearly needed improving. The slum dramatically exhibited crucial patterns of change and promised to shed light on the workings of urbanization throughout the city.

Among the aspects of disorganization appearing in the slum which intrigued Park and his students, juvenile delinquency aroused the greatest concern. Some of the most important research on this topic was done by Clifford Shaw and his associates. Shaw became interested in juvenile delinquency as a result of sociology courses at the University of Chicago and began a professional career as probation officer in the city courts shortly after World War I. The prevailing theory about delinquency at that time was that it stemmed largely from the "broken home" and social workers tended to find some kind of family disturbance in the background of every delinquent they encountered. Shaw became dissatisfied with their theory and the self-sealing mode of verification employed. If any family situation which had produced a delinquent was to be classified after-the-fact as a "broken home," whether it involved behavior as extreme as desertion or as mundane as an occasional quarrel, then this theory was not being satisfactorily tested. Moreover, in his own research, he began to notice certain patterns that could not possibly be explained by "broken homes." These disquieting observations revolved about the facts that delinquency cases seemed to concentrate in small areas and that active delinquents were involved in networks of gang relationships which apparently influenced their behavior to a considerable degree, even to the point of determining the types of offense they committed.

In 1923 Shaw was appointed research sociologist at the Institute for Juvenile Research in Chicago and because of this broadened perspective was able to develop a new interpretation of delinquency. The main element in this interpretation was the hypothesis that most delinquent behavior was a group activity, and that therefore the existence of delinquent gangs must be an important condition for its appearance. If a family lived in a neighborhood where adolescent gangs had developed patterns of delinquent behavior, then the likelihood that any normal boy would get involved in such activity was considerable. Of course, broken homes did accentuate the need for gang ties and removed any strong familial constraints against gang membership.

Shaw became very interested in the spatial distribution of delinquency. Using some of Burgess' hypotheses about urban growth patterns, he developed ecological spot maps showing the location of concentrated delinquency. The general model of five concentric zones proved applicable: Shaw was able to show that delinquency rates varied inversely in proportion to the distance of a locality from the center of the city. Today this finding that a delinquency gradient existed does not seem as important as some of the specific findings regarding neighborhood influences on delinquency patterns.

The central sociological problem is to explain the way in which the social organization of the slum affects family patterns and the formation of gangs in such a fashion as to promote delinquency. The exploration of this relationship is still in progress, though Shaw's early comments provide important leads:

"Persons live and act in families, clubs, schools, playgroups and gangs. These groups reflect community life, and the community itself reflects larger cultural and social processes. Behavior of a delinquent may be in part a reflection of a family conflict which drives him into a gang in which delinquency is a traditional group pattern. The delinquent gang may reflect a disorganized community life or a community whose life is organized around delinquent patterns."[12]

". . . with the process of growth of the city the invasion of residential communities by business and industry causes a disintegration of the community as a unit of social control. This disorganization is intensified by the influx of foreign national and racial groups whose old cultural and social controls break down in the new cultural and racial situation of the city. In this state of social disorganization, community resistance is low. Delinquent and criminal patterns arise and are transmitted socially just as any other cultural and social pattern is transmitted. In time these delinquent patterns may become dominant and shape the attitudes and behavior of persons living in the area. Thus the section becomes an area of delinquency."[13]

Delinquency is thus interpreted as the product of a certain kind of sub-community that arises as cities grow. This sub-com-

[12] Shaw, Clifford, *Delinquency Areas*, The University of Chicago Press, Chicago, Illinois, 1929, p. 10. The discussion of Shaw was greatly aided by Stuart Rice's perceptive essay on Shaw's development which appeared in a volume edited by Rice, *Methods in Social Science*, The University of Chicago Press, Chicago, Illinois, 1931, pp. 549-565.
[13] *Ibid.*, p. 206.

munity, the slum, develops when residential areas are invaded by business and industry as well as by impoverished immigrant groups and is characterized by deteriorated housing and a set of typical practices including organized crime, desertion, political machines, and other items identified by the Chicago sociologists as part of the institutional cluster in this "natural area."

Following Park's directives for the analysis of "natural areas," we can now attempt a preliminary codification of the interpretation of juvenile delinquency in Chicago as advanced by Shaw and his associates. For our purposes, Shaw's own preoccupation with the ecological distribution of delinquency in the city will not be given a central place, since it simply forms one step in a sociological analysis. Even if one supposed that the gradient hypothesis applies to every city, it remains necessary to investigate the ways whereby persons who live in areas distributed according to the hypothesis become delinquents. Ecologists are quite legitimately concerned with the location of "natural areas" rather than their social structure. But, as Firey has shown in his *Land Use in Central Boston*, even location is susceptible to socio-cultural influences. The main sociological question as to the place of the slum in the processes of urban disorganization and reorganization requires exploration of the social life of the sub-community. Ecological investigations really interpret land usage patterns so as to show how central areas of the city become slums and, therefore, breed delinquents. Sociologists are more interested in the latter processes than the former.

Several dimensions in Park's approach to "natural areas" can be singled out:

1. *Disorganization pattern.* The analysis should always begin with a pattern of behavior deemed problematic and the perspective on the basis of which it is so regarded must be carefully specified. Since this would presumably always entail threats to the social order of the whole city, preliminary distinctions must be carefully drawn between the extent of the

reality of the threat and its perception as such by different inhabitants with differing value systems.

2. *Distribution.* Here ecological and demographic techniques must be employed. The distribution of the disorganization pattern among various population elements should be determined so that areas having disproportionately high and low rates can be identified for further study.

3. *Sub-community social structure.* The way in which the disorganization pattern is viewed by the sub-community in which it occurs must be carefully explored so as to discover its relationship to other institutional patterns and sub-cultural values. Special attention must be paid to the way in which the pattern is transmitted.

4. *Urbanization.* The changing structure of the whole sub-community must be studied insofar as it has been shaped by the growth patterns of the entire city. This would include investigation of ecological processes, such as invasion and succession, along with more purely sociological data on the changing social and cultural structure.

5. *Reorganization.* Emerging mechanisms through which the city as a whole tries to suppress or contain the problematic behavior should be noted and their interplay with the institutional structure of the sub-community carefully explored.

This outline summarizes an approach to "natural areas" which emphasizes dynamic processes in the city. It would be possible to start with the "natural area" itself and examine its relationship to the whole city; however, there are advantages in using some form of disorganization as the point of departure. Beginning with a problem means that the inquiry will be led into exploration of the conditions for its existence. The "natural area" then is always brought into relationship with the processes of urban disorganization and reorganization.

We may now look at some of the specific studies undertaken by Chicago sociologists to examine them from the perspective established by our outline. The outline should prove helpful in underscoring aspects of the studies not ordinarily attended to. Our purpose is to examine ways in which "natural areas" have been studied in order to identify elements that must be considered and to locate points of departure for conducting new studies in modern cities.

Two of Shaw's books, *Delinquency Areas* and *Juvenile Delinquency in Urban Areas*, tend to be weighted heavily in an ecological direction. One supposes that this was a result of his early success at showing the existence of a delinquency gradient in Chicago. In any event, these books are primarily devoted to establishing the existence of similar patterns in other American cities. The success of this enterprise is considerable for the time periods and the cities included, but its generalized sociological significance is not carefully explored. Shaw presupposes the sociological underpinnings of his ecological findings when he assumes that the delinquency areas fall into similar positions because the process of urbanization in each of these cities has been identical with that taking place in Chicago. Similarly, when he assumes that the forms of delinquency and the "natural areas" in which they are bred must be like those of Chicago, he is taking too much for granted.

Shaw's other books, *The Jack-Roller* and *Brothers in Crime*, approach the problem more satisfactorily. Here, autobiographical material is used with great skillfulness to demonstrate the mechanisms whereby a slum socializes its residents to crime, and the personal disorganization consequent upon living in a socially disorganized area is vividly shown. Unfortunately, the focus is not on the social structure of the "natural area" except insofar as this impinges on and is perceived by the individuals writing the autobiographies. It is possible to reconstruct some conception of the sub-community social structure, but this is inevitably limited by the perspectives and biases of the informants. The whole conception of the relation between juvenile delinquency and urbanization could have been considerably deepened by research studying the whole "natural area."

[35]

A study like Thrasher's *The Gang* differs from Shaw's ecological as well as his life-history studies by starting with a slum institution that is central for juvenile delinquents. It is a kind of classificatory study dealing with a great many gangs but never going into any considerable detail on any one of them. Thrasher had data on over 1,300 gangs in the city of Chicago. He shows their distribution throughout the slums of the city and analyzes the processes of gang formation, functioning, and dissolution. Gangs are seen as mechanisms whereby slum children and adolescents achieve satisfactions not otherwise accessible in their underprivileged environments. His picture of gang activities and the satisfactions obtained therein is exceptionally vivid and probably still applicable to many contemporary slums.

In discussing the social structure of the gang, Thrasher provides a good many clues to patterns of control and leadership. The existence of similar patterns today, including distinctive names, codes, and territorial claims, can still be seen. His discussion of the ways in which gangs inculcate criminal attitudes supplements Shaw's life-history data on this same theme. The relations between younger gangs and their adult counterparts are shown so that the likelihood of transition from one to the other can be seen. Thrasher also treats the political role played by the gang and discusses some efforts at controlling gangs in Chicago.

Taken together, Shaw's and Thrasher's contributions give us a detailed picture of slum life in Chicago expressed through statistics, life histories, and non-systematic observations. Perhaps because of their efforts to deal with all of the slums and all of the gangs, they never get close enough to any single slum or gang. One cannot help but wish that they had chosen to focus on the social organization of one or more slums as they changed during the urbanization process. In this case the life histories might have been fitted into objective social structures rather than remaining isolated reports, and their cumulative contribution to sociology greatly enhanced. If Shaw and Thrasher had followed Park's directive to study "natural areas," they would have supplemented their studies with at least one detailed investigation of the effects of urbanization on the social structure of one Chicago slum. Juvenile delinquency would be seen along with gang life as part of the institutional configuration of the neighborhood. Its relation to family patterns, the church, the rackets, the police, and other significant institutions as this was affected by population influxes and cultural contact and conflict have to be explored in considerable detail. On the basis of this dynamic study of a changing sub-community, patterns might have been discovered that could have been traced in other slums so that some generalized conceptions as to the impact of urbanization could have been developed. Because Shaw and Thrasher pay little attention to the distinctive elements of the sub-communities in which the persons writing their life histories or the gangs studied were enmeshed, the advantages of a holistic approach to "natural areas" are lost. In similar fashion, because they assume the typicality of Chicago urbanization, they miss the possibility of seeing the effects of slower population influxes or more homogenous influxes on the formation of slum sub-communities.

A study of slums which comes much closer to satisfying this criterion is Zorbaugh's classic *The Gold Coast and the Slum*. It remains the best description of a complicated urban neighborhood and of its various sub-communities. Zorbaugh's main interest was in showing that a single section in Chicago—the lower North Side—could encompass several "natural areas" in close physical proximity but utter social isolation. The significance of this lies in the novel problem of social organization that arises when such disparate groups must be coordinated in terms of presumed common interests. If locality is to be the basis for urban political participation, this kind of neighborhood presents great problems. The effort at organizing a North Side community council so that the area could function as a political unit is described and the conditions for its failure analyzed.

This book is probably the richest source of information that we have about the "natural areas" in Chicago. It describes a slum, a Gold Coast, a rooming-house area, a Bohemia, and a Hobohemia or Skid Row. The institutional configurations in each are vividly depicted and their functional interconnections discussed. Thus, the world of furnished rooms is linked to taxi dance halls, all-night restaurants, burlesque houses, and suicide. The transition from boarding houses to rooming houses is described and its meaning as part of the urbanization process suggested. The paradox of social isolation and physical proximity is here carried to its farthest possible point: complete anonymity.

Zorbaugh provides similar pictures of the other "natural areas," using statistical material and personal documents to elucidate the characteristic social structures. He views the slum as the most disorganized region and is especially emphatic on this point with respect to Little Sicily, the Italian ghetto in Chicago. He sees the obvious symptoms—criminality, family breakups, political bosses, etc. One feels, however, that he never penetrates beneath the surface to see the distinctive social organization that is necessarily emerging in the slum, but here it may be that one's critical view rests on the advantage of hindsight and the brilliant research of William Foote Whyte.

The Gold Coast and the Slum is really the starting point for any further studies of "natural areas." However, it must be seen in a historical frame of reference before it can be so used. Zorbaugh's clues as to the social organization of each of the "natural areas" refer to that organization as it was shaped by the rapid population influx experienced in Chicago during the first part of the century. The social distance between the areas and the very distinctiveness of the areas themselves may well be a product of that historical experience. As a result, contemporary urban sociologists cannot mechanically apply this model but must use it as an exploratory instrument. Obviously, the extent of its application will differ from city to city, according to such variables as population size, heterogeneity, and special developmental experiences. Bohemias have all but disappeared, though one suspects that functional equivalents may be found, while the anonymity of the rooming-house district seems to have seeped outward to other areas on a large scale. Zorbaugh's definition of the problems involved in forming variegated neighborhoods into coherent political units remains pertinent though new factors must be taken into account.

Another extremely important study of a "natural area" is Wirth's book *The Ghetto*. The requirement that close attention be paid to a specific sub-community in terms of the effects of urbanization on its social organization is fulfilled by this study. Jews are distinguished from other ethnic groups in terms of their habituation to segregated living based on long experiences of this kind in Europe and other parts of the world. Wirth employs historical material on the establishment of European ghettos and the ways whereby they provided institutional shelter in a hostile environment. This strongly and fruitfully influences his interpretation of the first-generation American ghetto in Chicago. He shows how the ideologies, social patterns, and personality types of the ghetto provide a basis for ethnic solidarity under almost any circumstances; yet he is well aware of the tendencies toward emancipation in Europe as well as in America. With lessened emphases on the observance of religious ritual, the synagogue, Hebrew schools, kosher butcher shop and dairy—each so necessary for orthodox Jewish life— can be modified or abandoned so that the distinctive institutional structure of the ghetto tends to disappear.

In depicting the historical process of Jewish settlement in America and in Chicago, Wirth differentiates between German, Russian, and Polish Jewish ghettos so as to identify the distinctive social patterns of each. A cycle is noted wherein, during the early stages, the Jewish community is scarcely distinguish-

[39]

able from the rest of the city; but as urbanization proceeds, ghetto communal organization appears, then begins to differentiate sharply and finally disintegrates. He shows how the synagogue defends itself against American influences by modifying orthodoxy in the direction of conservatism. The conflicts within the community on this issue are graphically pictured as are the conflicts arising from Zionism. Personality types and vocational types adapting ghetto life to the American milieu are identified. The role of kin-village organizations called "Landsleit" is taken up though it is not quite as central in the organization of Jewish ghettos as it appears to be in Italian or Polish ghettos.

An important section of Wirth's book is devoted to the area of second settlement—that is, the neighborhoods to which ghetto dwellers move when they wish to live in a more American milieu. Jewish folkways are modified and often abandoned in favor of Americanized substitutes. Conflict between parents and children, as in the ghetto, arises around the greater Americanization of the children and their rejection of Old Country customs. The synagogue and religious life is less important but still finds some vigorous defenders.

During the twenties in Chicago, there was also some evidence that areas of third settlement were emerging in which distinctive Jewish customs would be almost entirely abandoned. This was the home of the reform synagogue, if any synagogue was maintained at all. Going a step farther, some Jews fled to the anonymity of hotel life. At the same time, Wirth does notice a tendency to return to diluted Jewish identification patterns on the part of some second-generation adults who had previously abandoned them.

Of all the Chicago studies, this book comes closest to fitting the outline presented earlier for analysis of social disorganization in "natural areas." It shows how the conditions of Jewish life in America led to the establishment, the disestablishment, and eventually, perhaps, the reestablishment of a different kind of segregated sub-community. We now suspect that the trend toward return to the ghetto has continued in the form of sec-ularized Jewish sub-communities differing only slightly from their Christian counterparts. Though Wirth's study does pro-vide a basis for interpreting changing patterns of ethnic life as they developed during the twenties in Chicago, it would have been valuable if he had compared these with Jewish life in small towns during the same period. The failure to see that Chicago was a special case in the analysis of urbanization pre-vented him from achieving depth in this direction equivalent to his depth in historical analysis.

Along with the studies of "natural areas" like the ghetto, there are a set of studies dealing directly with urban disorgani-zation and its symptoms. Among the best known of these is Faris and Dunham's *Mental Disorders in Urban Areas.* This research showed that certain kinds of mental disorders tended to be more frequent in selected urban areas than in others and advanced a set of hypotheses as to the relation between the sub-community milieu and mental disorders. The finding that paranoid schizophrenia was concentrated in the rooming-house districts of the city is perhaps the most significant, although their other correlations also deserve attention. Their hypothesis that the social isolation of the rooming-house district is con-ducive to schizophrenic withdrawal is an effort to relate social structure to personality functioning and to show how the in-stitutional patterns of a sub-community shape life organization. Following through the outline, they have identified a kind of disorganization (mental disorder), examined its distribution in the city so as to link its diverse forms to specific "natural areas," and then have begun to explore the ways whereby urbanization disorganizes these sub-communities. Criticisms of their research which point to the possible role of selective factors in attracting potential schizophrenics to areas of this kind do not vitiate the significance of their findings even though they suggest the need for more complicated interpretation. The direction for further research is closer examination of sub-com-

munities in which social isolation is common, so as to discern
its differential meanings and impacts on people with different
backgrounds. The larger implication that American cities increasingly isolate individuals and families so that this may become a generalized form of disorganization found in many
different sub-communities is also of the utmost importance.
If the assumption is made that communication is essential for
normal mental development, and large numbers of people can
be shown to lack opportunities for such communication in the
sub-communities in which they live, the likelihood of widespread mental disorders seems considerable.

A closely related study is Mowrer's *Family Disorganization*,
which examines desertion and divorce in Chicago during the
twenties. The same approach is employed, so that the relation
between "natural areas" and desertion or divorce is charted
and a correlation between high divorce rates and more well-
to-do areas as against desertion in less well-to-do areas is established. Both kinds of family disorganization seem to be increasing, and Mowrer interprets this as the result of urbanization which opens new opportunities to women, creates a sense
of restlessness, weakens primary group controls, and upholds
exaggerated versions of romantic love. Mowrer uses case-study
methods to illustrate difference sources of incompatibility in
the urban family along sexual, cultural, and other lines. He
distinguishes tensions due to incompatibility and response, economic individualization, cultural differentiation, and individuation of life patterns.

In addition to studies directly focused on types of urban
disorganization, Park directed his students' attention to special
occupations that seemed to reflect the peculiar characteristics
of city life. Some of these occupational systems such as the taxi
dance hall, studied by Paul Cressey, were of special interest
because they seemed to reflect urban disorganization. Others
like Donovan's studies of the waitress or Hayner's studies of
hotel life dealt with occupations and modes of life distinguished

by their contrast to rural patterns. In the case of the waitress we have functions typically allocated to the family in a small town being patterned on an impersonal basis in the city so that the preparation and serving of food becomes a profitable business. Hotel life blatantly displays urban impersonality and anonymity insofar as the closeness of physical contact that it entails renders preservation of maximum social distance imperative if privacy is to be protected. All of these studies link the occupation to various urban sub-communities that depend on it most heavily and analyze it in terms of its functions and dys-functions for these special sub-communities and the whole city.

Still another type of study is that completed by Everett Stone-quist in *The Marginal Man*. A specifically urban variant of the human condition, "marginality" is examined in all of its rami-fications: cultural, psychological, and sociological. In Chicago during the twenties, everyone (even the members of the most protected ghettos) experienced some marginality. Insofar as the secondary agencies of the large society permeated these "natural areas," the residents were exposed to different values and felt the conflicts that this engenders.

If we look at the city of the twenties from the perspective of the city of the fifties, the widespread "marginality" caused by exposure to diverse sub-cultures appears almost attractive when compared with the superficial homogeneity of exposure and response that characterizes much of modern city life. Perhaps the problem today is one of over-organization rather than un-der-organization, as was the case in the twenties. However, it must be recognized, as Park saw clearly, that these over-or-ganized communities fail to allow opportunities for expression of human differences and leave unsatisfied the inclination for status within a real community, offering instead only perpetual prestige-hunger which is whetted by the lack of real intimacy.

It is likely that the problems of the twenties—that of mar-ginality and coordinating diverse sub-communities—are no longer central. One suspects that secondary agencies like standardized news, entertainment, and education have leveled the population, reducing sub-cultural distinctiveness considerably. In the pendular movement from over-integration to over-individuation, the balance seems to have swung from the latter to the former. This is reflected in the current concern over "conformism" as well as by the feeble efforts at restoring ethnic customs and values in contexts where their meaning can no longer be revived.

Park perceived this likelihood too. His emphasis on communication as a mode of individuation rests upon appreciation of the importance of significant communal ties: "In the little worlds where people come close together, human nature develops. The family, the tribe, the local community are instances. To the extent that we have intimate relations, we get responsiveness to one another. The definite personalities that we know grow up in intimate groups. In the larger society we get etiquette, urbanity, sophistication, finish. Urbanity is a charming quality but it is not a virtue. We don't ever really get to know the urbane person and hence never know when to trust him. It is more or less fundamental traits of personality which arise in the intimate group which enable us to act with definiteness and assurance towards others. Manners are of secondary importance."[14]

This is no simple idealization of the primary group but rather an effort at defining its role in human individuation. He is at least as aware of the need for privacy: "We not only have, each of us, our private experiences, but we are acutely conscious of them, and much concerned to protect them from invasion and misinterpretation. Our self-consciousness is just our consciousness of these individual differences of experience, together with a sense of their ultimate incommunicability. . . . For a certain isolation and a certain resistance to social influences and social suggestion is just as much a condition of sound

[14] Park, Robert, *Race and Culture*, The Free Press, Glencoe, Illinois, 1950, p. 22.

personal existence as of a wholesome society. It is just as inconceivable that we should have persons without privacy as it is that we should have society without persons."[15]

But the later stages of urbanization provide neither the opportunity for individuation through learning about oneself and others during intimate interchanges, nor the opportunity for individuation through privacy and resistance against social influences. Instead, standardizing social influences have become more subtle and situations requiring the wearing of "social masks" that prohibit real intimacy have become more pervasive. The full implications of Park's social psychological comments will be taken up in Part III, where the effects of suburbia on personal identity are discussed.

Park's theory focuses on the pressures that urbanization exerts on communities affected by it. It relates these pressures to problems of control and change in the city so as to elucidate the central structures arising during the transition. Where the population influx is rapid, immense, and heterogeneous, it is likely to sort itself out into the kinds of sub-communities identified by Park. Some of these sub-communities will probably develop deviant ways of life that should be controlled in the interests of the city as a whole. Secondary control agencies designed to suppress disorganization or nip it in the bud by guaranteeing effective socialization will appear. The *specific* conceptions about the varieties and functioning of "natural areas" and secondary controls, based as they are on the city of Chicago during the twenties, must indeed be *critically* employed. At a minimum, they do provide a preliminary hypothetical model against which the actual social structure and transitional processes in any community can be studied.

Further development of the theory requires greater specification as to the conditions under which urbanization occurs in other communities to establish more complex typologies, among

[15] Park, Robert, *Human Communities, op.cit.*, pp. 175-176.

which the Chicago pattern will become an important prototype. Effort must be made to extend the theory through time so that the characteristic effects of present-day population influxes on contemporary cities can be compared with the earlier stages of urbanization so carefully studied by Park. We remain interested in urbanization as a disorganizing force but must recognize that its nature as well as its effects will be different because the environing pressure, the urban influx, and the system (the responding community) differ considerably from their counterparts in the twenties. West Coast cities, for example, are growing very rapidly, but the incoming population has, for the most part, lived in cities before and is more Americanized. Thus, the problems arising out of their incorporation into city life are not the same as the problems created by foreign immigrants of peasant stock.

It is the task of the urban sociologist of the sixties to identify these new problems and interpret them in terms of a modified theory of "natural areas." This modified theory will have to include analyses of the social structures of urban sub-communities that have appeared since Park's research. It will also have to consider the structural transformations in the traditional "natural areas" occurring since they were studied by him. Part II of this book takes up three urban sub-communities: the slum, Bohemia, and the suburbs so as to show the patterns of disorganization and reorganization appearing in them. Park's theories about urban social processes must indeed be brought up-to-date, but his keen curiosity about life in the city and his willingness to leave the confines of academic cloisters to go out and explore it can hardly be improved upon.

CHAPTER 2

THE LYNDS AND INDUSTRIALIZATION
IN MIDDLETOWN

A SET of studies comparable in depth and importance to those carried out by Park and his students are the two investigations of Muncie, Indiana, undertaken by Helen and Robert Lynd. Their full picture of life in a medium-sized American town during the mid-twenties and then during the mid-thirties has drawn the praise of non-professional as well as professional readers. In fact, much of our image of small-town life during these periods comes from the Lynds' work, supplemented by the astonishingly similar literary reports of Sinclair Lewis. Just as Theodore Dreiser provides novelistic documentation of the human dramas enacted in the growing cities around the turn of the century, thereby supplementing the interpretations of the Chicago sociologists, so there is an equally significant correspondence between *Main Street* and *Babbitt,* on the one hand, and *Middletown,* on the other.

It is all too easy to read both *Middletown* volumes as if they were purely descriptive reports. Indeed this is the fashion in which they are most often read. The mass of absorbing details lulls the reader into an aesthetic rather than a scientific frame of mind. But succumbing to this temptation diverts attention from the crucial sociological generalizations in both books. The first volume contains the main conceptions necessary for interpreting the effects of industrialization on American community life; the second examines the impact of the depression on the changes described in the earlier volume. Together they constitute an exemplary sociological description of two historical processes affecting communities throughout the country.

The problem that these volumes present to the theorist is quite different from that presented by Park. Where he never hesitated to generalize about the character of urbanization from his observations in Chicago, the Lynds, with equal consistency, refuse to draw generalized conclusions about industrialization or the effects of the depression from their research in Muncie. The Chicago studies had to be placed in their specific socio-historic context so that relevance to other contexts could be perceived. With the Middletown series, the task involves discovering directions of potential generality amid the mass of reported details.

There is no "easy" solution, such as turning to the last chapters for analytic summaries, since the Lynds characteristically keep their concluding chapters very short and not very much more generalized than those preceding them. Luckily, they do include generalized formulations in chapters dealing with substantive aspects of community life; yet these are never drawn together. For that matter, while it is evident that the second volume focuses on the depression, it is really never entirely clear whether the authors set out to study industrialization in the first book. The importance of industrialization probably grew clearer as the research progressed, although they must have been interested in it from the start.

Thus their initial decision to study Muncie involved a desire to explore a community which would "be as representative as possible of American life" and at the same time "compact and homogenous enough to be manageable." The former criterion was interpreted as demanding that there be "an industrial culture with modern high-speed machine production" (among other requirements listed on page 7 of *Middletown*). Another clue to their interest in industrialization was their choice of time span for the study, 1890 to 1924. Chapter Three of the first volume makes it quite clear that the "gas boom" in 1890 provided the initial stimulus for industrial development and their comment that "this narrow strip of thirty-five years comprehends for hundreds of American communities the industrial revolution that has descended upon villages and town . . ." expresses this interest quite explicitly.

In choosing the community for study, they showed considerable ingenuity by eliminating the effects of urbanization: "In a difficult study of this sort it seemed a distinct advantage to deal with a homogeneous, native-born population, even though such a population is unusual in an American industrial city. Thus, instead of being forced to handle two major variables, racial change and cultural change, the field staff was enabled to concentrate upon cultural change. The study thus became one of the interplay of a relatively constant native American stock and its changing environment. As such it may possibly afford a base-line group against which the process of social change in the type of community that includes different racial backgrounds may be studied by future workers."[1]

Though the town did grow from 11,000 people in 1890 to 35,000 by 1924, the newcomers came largely from nearby farms rather than from foreign lands. Thus the disorganization often accompanying the assimilation of culturally distinctive groups was minimized and the formation of typical urban "ghettos" avoided. Muncie became a sort of social laboratory in which industrialization as a source of change could be isolated from urbanization.

The Lynds approached Muncie with the following conceptions:

1. Their main purpose was to study the effects of industrialization on community life.

2. This was accomplished by tracing these effects through six major areas of life classified according to W.H.R. Rivers' scheme as "getting a living, making a home, training the young, using leisure, engaging in religious practices, and engaging in community activities."

[1] Lynd, Robert S., and Lynd, Helen M., *Middletown*, Harcourt Brace & Co., New York, 1929, p. 8.

3. They assumed that changes in any one area would reverberate throughout the others to a greater or lesser degree.

4. The area that reverberated most widely was "getting a living" because it encompassed many of the central changes wrought by industrialization.

5. Distinctions in ways of "getting a living" filter the impact of industrialization so that two main groups affected differently by it were the working class, who "dealt primarily with things," and the business class, who "dealt primarily with people."

Such general orientations are most useful for organizing field data, but they are hardly suitable as a theoretical summation of findings. The task of pulling such a theory out of these volumes is more formidable.

Some perspective on the social changes in progress during the period covered by the first volume can be attained by glancing at a summary comment on the conditions necessary before industrialization can occur in "backward" countries advanced by Wilbert Moore in his useful book *Industrialization and Labor*: "What is the most general explanation that may be given (for) the pervasive strength of the industrial system in subverting social systems and surmounting or penetrating the natural barriers that a balanced nonindustrial system possesses? . . . A tentative generalization may be hazarded; it is the positive, institutionally sanctioned, and structurally necessary prescription of *mobility* that is at once the source of productive efficiency of the industrial system and the source of disruption of nonindustrial systems. The beginnings of the institutionalization of mobility are fraught with difficulties almost as various as the societies encountered."[2]

It is exactly those "beginnings" and "difficulties" that the Lynds deal with. Considerable historical imagination is required to conceive of the possibility that American community life was once badly strained by industrialization; yet when

[2] Moore, Wilbert E., *Industrialization and Labor*, Cornell University Press, Ithaca, New York, 1951, pp. 311-312.

this imaginative leap has been accomplished some important conceptual tools can be developed for interpreting present-day strains arising from the same source.

Moore's condition, the institutionalization of mobility, was actually met by Muncie long before 1924. As the Lynds show, however, it is not the simple orientation to mobility that matters, but rather the orientation toward mass-produced commodities as symbols of mobility coupled with willingness to perform specialized occupational roles in order to be able to purchase these commodities, that constitutes the mainspring of the American industrial community. Here, mass production and mass consumption appear as interrelated phases in a recurrent cycle so that motivation to perform roles demanded by the former is maintained by providing access to goods which are prerequisities for desired life styles. Unlike many "backward" countries, even pre-industrial Americans were concerned with mobility, but the character of this concern differed sharply from that exhibited during the post-industrial period.

The Lynds show that the fulcrum of all the changes taking place in Muncie between 1890 and 1924 was the breakdown of the craft hierarchy occasioned by the introduction of machines and assembly-line techniques in local factories. This technological transformation wrought a social transformation which, if properly comprehended, provides a thread winding through all the specific institutional changes reported in *Middletown*. The social organization of Muncie in 1890 included the same two groups found in 1924, the business class and the working class. But their relationship and relative positions were quite different. The craft workers were as highly respected, politically influential, and well paid as their business counterparts. They were better organized than the businessmen, who could only envy the effectiveness of the craft unions in providing economic and social facilities for their members. Opportunities for social mobility as a craftsman were plentiful, and reasonable application almost guaranteed that anyone starting as an apprentice would, in due time, work up to the position of master craftsman. They were actually more certain of succeeding than their fellow townsmen who went into business, since craftsmen avoided the uncertainties of entrepreneurship.

Between 1890 and 1924 industrialization shattered the occupational basis for craft employment by rendering skilled laborers unnecessary for most jobs. The Muncie craftsmen found themselves on a level with and even at a disadvantage in competition against younger unskilled members of the labor force. Their craft associations declined as the occupations on which they depended dwindled. Where craftsmen had formerly tended to live close to their places of employment, the development of transportation systems and the loss of attachment to their jobs meant that they could distribute themselves throughout the city. As residence patterns shifted in this fashion, working-class neighborhoods lost their cohesiveness and "visiting" declined while the number of socially "isolated" families increased. Workers who wished to maintain their own self-respect and the respect of their families were forced into the scramble for household gadgets. Their children demanded educational opportunities and this quickly became a major goal for working-class parents, since they saw in it a way of helping their children to regain the status they themselves had lost. Drastic changes in working-class family life, including acceptance of the need for wives and mothers to take jobs, were justified in terms of the new status requirements.

Meanwhile the business class had far fewer changes to make. It was already accustomed to impersonal status display and welcomed the flood of commodities that industrial technology made available, just as it welcomed the flood of new entrepreneurial and managerial opportunities that were opening up. Their clubs—the Boosters, the Rotary, the Lions, the Kiwanis, and other familiar organizations of the American small town—assumed community leadership as the craft unions collapsed. These newly risen business leaders saw their way of life—

mobility symbolized by acquisition and display of commodities—become the dominant if not the only basis for placement in the Muncie social structure.

After elucidating this fundamental transformation in status structure as a result of which working-class craft mobility was eliminated and business-class mobility models substituted, the Lynds describe the processes by which working-class life was reorganized to equip its members for emulating these models. Every institutional area—the home, the church, the school, recreation, and politics—accommodated to the requirements imposed by industrial work roles as well as to the mass-produced commodities flooding the market. New institutions such as advertising appeared to stimulate demand in areas where manufacturers considered it to be lagging. All of this emanated from outside of Muncie so that the sense of local autonomy declined along with the sense of individual autonomy characteristic of life in 1890.

Each specific change in life style described by the Lynds can be interpreted analytically as a response to industrialization. Such interpretation, however, involves keeping in mind the inclusive status reorderings through which industrialization is realized in this particular community. Changes in working-class patterns rest upon status insecurities caused by shifts in their community positions as a result of the breakdown of the craft hierarchy. Their efforts at imitating middle-class standards of living arise out of the destruction of their previous standards and the associational ties which these had provided. Business-class leisure activities begin to revolve around business contacts, but workers, denied this possibility, use leisure to compensate for and escape from their meaningless job routines. Both groups depend increasingly on mass-produced standardized facilities like radios, movies, and automobiles. The schools are pressed into service for training girls to manage mechanized households and boys to assume mechanized jobs. But the school does remain one place in which class background is suspended in at least one realm, high-school athletics. Here, the mobile working-class boy with exceptional gifts can gain access to middle-class groups. The reader has to relate the myriad of specific changes in Middletown life between 1890 and 1924 to the major status reorderings that accompany industrialization. Until he does this, the problem of sorting out the changes specific to Muncie from those having more generalized relevance can hardly be dealt with.

There is no question but that some of the experiences of the two classes in Muncie during this interval were rather unique. The Lynds deliberately chose a community in which this would be the case. Probably as many American workers encountered the impact of industrialization within the confines of urban ghettos of small towns, and their experiences must have differed significantly. But the essential fact remained constant—that life plans, whatever they may have been, had to be reoriented around the pursuit of status through money and commodity display. Thus, the willingness that the second generation of urban Americans of every nationality showed to move to "more modern" areas of second settlement rested upon this commitment. And the institutional reorderings involving the abandonment or dilution of distinctive ethnic life styles in the second settlement, paralleled the changes in working-class life styles reported by the Lynds in Muncie. Immigrant Polish peasants, as Thomas' and Znanecki's *The Polish Peasant in America* shows, started with rather different values from those held by Muncie craftsmen in 1890, but the processes of shedding these values to adopt American middle-class standards were no less painful and the resulting style of life not too different. Americanization in the sense of accommodation to industrialism, which, incidentally, was clearly one of the main senses of the term during the twenties, proceeded in a similar fashion among urban peasant immigrants and small-town craft workers. The immigrants may have had to make a sharper transition, but at least they had the advantage of knowing that they were being required to change, while the Muncie workers had little sense of the collective character of their experience, nor could they come to terms with it as part of the necessary accommodation arising from a shift in homeland.

Urbanization and industrialization are obviously deeply interwoven during the twenties. Indeed, Park frequently refers to industrialization as a major source of social change. Cities required an industrial technology to sustain their populations while factories needed the concentrated labor forces that cities provided. But it can be fruitful to ignore temporarily the interrelationships so that the distinctive features of each can be examined.

It is this task of disentangling the element of industrialization which distinguishes the first Middletown volume. Furthermore, from the standpoint of a sociological case study, the Lynds' methodology is superior to any adopted by the Chicago sociologists. By studying Muncie in 1924 against a base-line of 1890 they provide delineation of the pre-industrial community with a detail and clarity entirely lacking in our conceptions of pre-urban Chicago. All too often one finds the Chicago sociologists employing an almost stereotyped image of rural life to juxtapose against their conception of urban conditions. Park rarely offends in this fashion, but his students are not always as careful. Obviously, gathering data on the development of urbanization in Chicago from its rural origins would have been a far more massive task than tracing the development of Muncie from a craft to an industrial community. Furthermore, much of the prior experience of Chicago residents during the twenties had occurred in foreign lands. So Wirth's fine description of ghetto life is distinguished by its careful exploration of Jewish community life in Europe as a background for understanding the social life and problems of the areas of first settlement. But the requirement, met so well by the Lynds, that a case study of a community in transition must

[55]

describe the social structure prior to the onset of the changes is entirely valid. Chapter 4 will take up this problem in further detail, but the methodological point is simply that studies of communities in transition should be explicit about the structure from which the transition starts (i.e., rural Chicago, craft-centered Muncie) so that effects of this same transition in communities with different starting points can be studied.

The same careful attention to what Park might well have called "natural history" marks the second Middletown volume. Indeed, *Middletown in Transition* is even more of a methodological coup. The Lynds were not simply repeating their earlier study, nor were they examining the effects of the depression on community life. The unusual theoretical significance of their research stems from the explicit orientation to developmental stages that dominates the second volume. It is directly in line with the first book in that it focuses on the effects of the depression on the changes set in motion by industrialization as reported in *Middletown*. Thus, we have a sociological history of industrialization in this community from 1890 to 1935. The second not only adds the time segment, 1925-1935, but, more importantly, it allows us to see how a historical event, one dominant event of the decade of the thirties, affected the main directions of change in a community where these had already been carefully plotted. Since this particular historical event was by its nature closely linked to the main drift, the conjunction of research opportunities is exceptionally fortunate.

Middletown in Transition starts from the problem, "How did the depression, with its concomitant denial of access to the new sources of industrial status observed earlier, affect the commitment to these sources of status on which industrialization presumably rests?" It was the special status-deprivations, so closely linked to the earlier anxious acceptance of industrial mobility, that made the depression a peculiar threat to the life order of Muncie. Since this was the emerging American life order, findings about factors affecting its character in this community might well provide clues as to its effects in other quite different communities. Again, the Lynds do not tie together their observations in any general summary, but this is not quite as hard to do as was the case in the first volume. Here we have the advantage of knowing the main directions of change resulting from industrialization, so that our anchoring in the historical flux is secure.

Perhaps the most general finding of the second volume was that the depression, though it shattered the immediate life plans of Middletowners, was viewed pretty consistently as an "interruption" to be patiently "waited out" rather than as a symptom of irremediable social breakdown. Old ideological slogans, like "Onward and upward with free enterprise," were clearly not borne out in social experience, but perception of the disparity was blurred by the zigzag process of the depression, which was interpreted by the business class as proof that it was nothing more than a "dip" from which recovery was imminent. Furthermore, both the workers and the businessmen of Muncie were protected from the worst ravages of the depression by the fact that the main industry of the town was the production of glass jars, one of the few commodities to remain a household necessity during the depression.

This peculiar fact meant that Muncie was on the whole better off than neighboring towns. The owners of the glass-jar factory, the X family, made quite certain that no one was left unaware of this fact. During the depression, they consolidated their hold on the economy as well as the social and political life of Muncie. The following comment by a local citizen in 1935 is quite revealing: "If I'm out of work I go to the X plant; if I need money I go to the X bank, and if they don't like me I don't get it; my children go to the X college; when I get sick I go to the X hospital; I buy a building lot or house in an X subdivision; my wife goes downtown to buy clothes at the X department store; if my dog strays away he is put in the X pound; I buy X milk; I drink X beer, vote for X political parties,

[57]

and get help from X charities; my boy goes to the X YMCA and my girl to their YWCA; I listen to the word of God in X-subsidized churches; if I'm a Mason I go to the X Masonic Temple; I read the news from the X morning newspaper; and, if I am rich enough, I travel via the X airport."[3] The Lynds proceed to document this objectively by tracing out the ramifications of the X family power in considerable detail.

There is some question as to whether the Lynds overlooked X family power in their first investigation or whether this power position was not sufficiently significant in 1924 to justify separating the X family from the rest of the business class. Perhaps another factor, the growing interest in Marxism among American intellectuals as a whole and on the part of the Lynds in particular, led to identification of this "ruling" family in Muncie. Certainly their treatment of power in 1935 shows signs of Marxist influence. They are concerned with showing how the wealthy and powerful families, using every agency at their command, protect their class interests. Trade unions trying to organize the city are held at bay by the X family while public expressions of optimism are encouraged. Muncie workers and businessmen are propagandized to consider themselves "fortunate" because some of their industries remain open as compared to "union" towns, which are shut down completely. The schools, the newspapers, and the churches are all brought into line, spreading the doctrine of loyalty to "capitalism, Americanism and Middletown."

The Lynds make an important theoretical step forward from their earlier treatment of power. They force attention to the possibility that systems of social control can operate in the interests of one group and against the interests of others. This is an advance beyond Park's assumption that controls serve the interests of the whole community, though there is evidence from Park's specific analyses of problems that he was well

[3] Lynd, Robert S., and Lynd, Helen M., *Middletown in Transition*, Harcourt Brace & Co., New York, 1937, p. 74.

aware of this other possibility. By revising their conception of Middletown social structure to include the X family at the top of a hierarchy, the Lynds link economic class to the distribution of power and influence more directly than did the urban sociologists. Yet similar power arrangements undoubtedly prevailed in the cities during this period, although one-family domination was probably confined to small towns, and to but a few of them at that.

There were many signs in 1936 that the power wielded by the X family in Muncie would decline in the near future. It seems likely that this family assumed a dominant position during the depression because they felt threatened and wished to preserve an acquiescent labor force. But even in Muncie, as the Lynds indicate in footnotes, industrial unionism was beginning to take hold. The large working-class vote for Roosevelt in 1936 was one sign of the change. Workers found that their avenues of mobility, already shrunken by industrialization, were being blocked still further as the depression eliminated the last vestiges of craft skills. During the pit of the depression they were too preoccupied with hanging onto the piecemeal jobs still open to them to worry about "getting ahead." Slight improvements around 1935, however, reawakened interest in the "dream," which had been dormant but never extinguished.

By then, however, the prospects of mobility through industrial jobs appeared even less promising than earlier. The readiness to accept trade unionism as a way of getting their "share of the pie" was beginning to spread. As the victories of the unions in urban centers were publicized, a movement to get on the bandwagon emerged even in such a tightly controlled town as Muncie. Simultaneous with this glimmering of working-class resurgence, the Lynds observed that, compared with their parents, the second generation of the X family displayed much less concern with Middletown affairs. Having been educated out-of-state, they tended to look eastward toward the big cities, so that their styles of life diverged sharply from those of this Midwest town. They were interested in state-wide power. Another wealthy Muncie family, the Y's, had always focused their attention on extra-local affairs, thereby setting a precedent for the younger X's to follow. Furthermore, the relative power of all the local "upper class" families must have declined somewhat when General Motors reopened its Muncie plant and brought in absentee power of considerable scope.

While the chapter on the X family is the only one in *Middletown in Transition* having no precedent in *Middletown*, the chapter on "caring for the unable" is the one that underwent the greatest expansion from the first volume to the second. This was clearly the area of major institutional reorganization since most working-class and some middle-class families suddenly found themselves involuntarily thrust into the "unable" category. Here was the cutting edge for the erosion of traditional conceptions of "individualism, self-help, local autonomy" and many similar values. Community responsibility for the involuntarily unemployed had to be accepted. Even more important, this was treated as their "just due" rather than as "charity." Men on relief resented being forced to take money but really had little choice if they wished to feed themselves and their families. Pride in "self-sufficiency" gave way before the problems of obtaining food, clothing, and shelter. This situation affected the whole community as local funds for relief purposes quickly ran out and left the town dependent on the federal government for further aid.

As in the first volume, *Middletown in Transition* contains a long, detailed chapter on changes in each area of life. Family life was badly shaken by the depression and the working-class family especially had to make major readjustments. Having an unemployed husband around the house disrupted family routines. Certainly the males were deeply threatened by the experience, and one suspects that their later willingness to join unions was reinforced by their desire for protection against intolerable family strains arising from unemployment.

Education, leisure, and religion became more diversified as the community grew in size and heterogeneity. Aspirations for more education for their children continued to soar among parents of all classes. Business-class control over the kind of ideas taught in schools increased as fears of "radicalism" spread. "Progressive" ideologies emphasizing "individual attention and self-expression" diffused to Middletown educators, but the realities involved in operating an overcrowded school system kept them from being institutionalized. The tendency toward decline in total family participation in common leisure activities noted in 1924 continued unabated. Radio, the movies, and automobiles retained and even strengthened their hold in spite of the increasing strain they placed on family budgets. Church attendance continued, though with some decline but, more significantly, the superficiality of religious participation persisted in spite of the presumed greater need for solace during the depression. Ministers remained on the periphery of local power circles while their standardized sermons became even more remote from the affairs of the day than they had been in 1924: ". . . going to church becomes a kind of moral life-insurance policy and one's children go to school and college so that they can get a better job and know the right people. Bit by bit in a culture devoted to movement and progress these permanent things of life become themselves adjuncts to the central business of getting ahead, dependent symbols about the central acquisitive symbols of the community's life. And so progress recaptures and confines its own children."[4]

Business values then continued to hold sway over the community, with other potential value-defining agencies being forced to accommodate themselves as they had already begun to do in 1924. After this research was completed, the trade unions provided a kind of alternative on the local scene. But it is doubtful that the unions ever challenged the main protocol of business philosophy, the merit of "getting ahead"; instead, they

[4] *Ibid.*, pp. 317-318.

simply offered a new formula to assist the workers in their upward climb.

Significant changes were reported during the depression in attitudes toward the federal government and its role with respect to local life. This was most clearly reflected in growing acceptance of federal relief funds and projects to help unemployed families and individuals survive. The depression itself focused attention on national affairs, since its origins as well as its solution had to be sought on that level. Interest in local politics declined accordingly, as attention shifted to the national scene. The press reflected this shift by devoting an increasing amount of space to national and international news while cutting local coverage. Syndicated material increased greatly. Despite efforts by the Muncie upper class to swing opinion and votes behind Landon in 1936, the Lynds report that Roosevelt carried most of the working class and even half of the middle class, thus giving the Democrats a decisive victory in this normally Republican community.

Middletown in Transition ends with two chapters summarizing the changes in Middletown life since 1924 which bring our picture of the social structure up-to-date, in this instance up to 1936. There is a detailed survey of Middletown values which shows that the depression had not made any profound change in the dominant ideals or norms except to qualify them at certain points in order to accommodate depression-imposed innovations like "relief" as necessary evils. Comparing their list with a similar list compiled by Lord Bryce thirty years earlier, the Lynds observe that there has been a shift in emphasis from adventurousness and new experience to security.

It is important to note that the cultural system described by the Lynds was that of small-town, white, Protestant Americans though it may have been shared to a greater or lesser degree by other social categories. More important than any substantive changes in the expressed ideals was the appearance of wider gaps between ideals and everyday realities between

1924 and 1936. In 1924, the spirit of "progressive optimism" was far easier to maintain than anytime during the depression. Similarly, confidence in an ever-improving future was severely shaken by the ever-worsening present. But the symbol creating agencies remained under the control of the business class: "This symbolic ceiling over Middletown is largely set and defined for the city nowadays by its business class. The chance for the mass of the population to 'go up in the world' to affluence and independence appears to be shrinking noticeably. It so happens, however, that those who still retain the best chance to rise in the world, to skim the cream from the economy, also control the press, the radio, the movies, and the other formal media of diffusion of attitudes and opinions. They are thus in a position—in the kind of urban world suggested above, containing an increasing group of untied residents who do not contribute materially to the native ideologies of the folk—to tell the city full of people largely living off the skimmed milk of the economy what to believe."[5]

Every effort was made to stretch the old concepts to fit the new, more complicated realities. Even the depression itself was treated in this fashion though the simplified formulae of the local press clearly could not gloss over all the real hardships that unemployment entailed. "Initiative," "independence," and other cherished virtues possessed little relevance when no jobs in which to exercise them were to be had.

Industrialization had already exposed Middletowners to the necessity for abandoning traditional ways of living in the factory as well as the home. It has widened the gap between religious ideologies emphasizing "brotherhood" and the competitive realities of economic life. By 1924 it had become common practice to accept any alterations in life style that the dominant image of "success" required. When doing so entailed behavior that contradicted traditional values, this was absorbed without conscious strain by segregating the realm of ideals from that

[5] *Ibid.*, p. 471.

of necessities. Everything involved with "getting ahead" was placed in the latter category. By 1929, when the depression struck, the people of Muncie found themselves with "getting ahead" as the only life goal that had any genuine meaning. In face of profound threats to its realization they clung to it even more desperately than before. But they were shocked into some awareness of their personal vulnerability.

"One suspects that for the first time in their lives many Middletown people have awakened, in the depression, from a sense of being at home in a familiar world to the shock of living as an atom in a universe dangerously too big and blindly out of hand . . .

"And the more the disparities have forced themselves to attention, the more things have seemed 'too big' and 'out of hand,' the more Middletown has inclined to heed the wisdom of sticking to one's private business and letting the uncomfortable 'big problems' alone save for a few encompassing familiar slogans."[6]

The response to social complexity, to being small cogs in large factories and communities, was anxiety followed by privatization. An increasingly abstract and attenuated relation to the larger social forces affecting job and community was accepted while attention focused on personal strategies for piloting oneself and one's family through the storm.

It is unfortunate that this withdrawal from larger perspectives occurred at a time when the demand for complex understanding and moral choice was heightened: "A too-little-considered aspect of our current culture is the extent to which in all sectors of living problems have grown in complexity far beyond their former proportions. It is less tolerable today than a century ago to subsume the problems of living under simple blanket formulas. As localism has dwindled and science and technology have invaded field after field, simple activities—all the way from nutrition and child rearing to making goods and

[6] *Ibid.*, pp. 491-492.

carrying on international relations—have become complex matters of defying even the experts at many points. 'Science,' the 'growth of knowledge,' and 'progress' have caused disturbing questions to sprout where fewer choices were recognized before. This has put at many points in our institutional life strains upon the human agents in the culture never anticipated when these institutional frameworks were erected in an earlier era."[7]

Abdication of responsibility in the face of complexity was fostered during the depression by increasing isolation of individual families with Middletown neighborhoods. As urbanization spread, individuals lost their sense of the whole community. The whole social structure, intelligible in 1924 in terms of a division between working and business classes, was beginning to show at least five distinct class groupings and considerable ethnic complication in 1936. "Success and the dollar" became the only symbols of community placement understood and accepted by all.

Middletown in Transition reports the final breakdown of any sense of the character or workings of the whole community among most of its residents. It was no longer possible to understand the processes of Middletown life in terms of the functioning of its business class and its working class. Outside forces intervened to disrupt the life activities of both. Any sense of control over their own affairs or over the affairs of the whole town was dissipated by the depression. Even the powerful X family could not stave off the depression completely. The new groups left the older Muncie workers and businessmen feeling rather like strangers in their own town. The only mechanisms for readjusting life activities locally involved accepting national organizations like trade unions and the Democratic party, though this could hardly restore the sense of community that had existed earlier.

Muncie workers, increasingly interested in security, were turning to trade unions only because local supports had failed

[7] *Ibid.*, pp. 380-381.

and not, as the Marxists might have expected, out of any profound sense of working-class solidarity. Federal relief had shown them that they could be helped in their depression-generated distress by political intervention. But this recognition never included the further commitment to any interpretation of the sources of their distress which located its causes in the capitalist system nor did it involve self-definition as members of an exploited class. Their loyalties to the Democratic party and eventually to their trade unions promised to take the same form as their loyalties to "Middletown, capitalism and progress." Surrounded by impersonal large-scale enterprises at work and at home, and badly frightened by their own vulnerability as revealed by the depression, they turned toward alternative large-scale enterprises, the unions and the political parties, only because these showed some promise of alleviating individual misfortunes.

In this view, the depression fostered the sense of alienation from communal and occupational structures that had begun with the destruction of the craft hierarchy by industrialization. Substitute organizations with which workers affiliated, like trade unions, never permeated their life activities or became sources of genuine status as had the old craft employments and craft unions. Instead, the new industrial unions and the Democratic party were often perceived as holding operations through which minimal places on the mobility ladder could be secured without requiring deep emotional commitments of any kind. Workers learned to distinguish large-scale enterprises which aided them from those that opposed their immediate interests. Even here they easily got lost in the sea of symbols and so were able to affirm loyalties to all of the large-scale enterprises impinging on them even though the enterprises, like unions and management, were often in conflict with each other. The same ideological vagueness that sheltered them from institutional contradictions created by industrialization helped them to function in new bureaucratized settings. But the price they paid was withdrawal of real loyalties or affective involvement from the organizations to which they belonged. To go into this further would anticipate the next chapter, but it is worth noting that working-class adaptations in Muncie during the depression crucially involved reactions to bureaucratization.

To summarize the foregoing comments on *Middletown* and *Middletown in Transition*, it is clear that the Lynds have provided a case study in depth of major changes in community patterns comparable to urbanization. Starting with industrialization, which is the implicit central theme of both volumes, they show the conditions accompanying the introduction of industrial production and consumption patterns in a Midwest town formerly dominated by a craft economy. Their research is distinguished by the fullness with which the social structure before the changeover is described. Because of this, the essential features of the transformation are highlighted in such a fashion as to facilitate separating them from the specific form it took on in this particular community. The breakdown of the craft hierarchy in the glass factory, with its concomitant shift in the relative statuses of the business class and working class, can be identified in any community in which a craft economy is subjected to reorganization using machines and assembly-line techniques to eliminate skill. Warner traces a parallel process in the Yankee City shoe industry, though complications enter that will be considered in the next chapter. The simultaneous reorganization of status placement on the basis of competitive acquisition of mass-produced goods observed in Muncie will always accompany mass production, though here too Yankee City displays interesting differences.

Middletown in Transition sets up a new model through which the effects of important historical events marking the experiences of a decade can be fitted into explorations of basic community processes. In this case, the dominance of the X family colored responses to the depression by both classes in Muncie. Finally, however, trends that could be anticipated in

[67]

other communities began to manifest themselves here. The responses of the Muncie workers to trade unionism may have involved more resistance than urban industrial workers displayed, but their eventual acceptance exposes some of the deeper attitudinal underpinnings shared by most American workers. The unions and the Democratic party were accepted as counter-bureaucracies necessary for self-protection but hardly valuable in themselves. "Getting ahead" as a central value survived the objective setbacks experienced by most citizens in Muncie during the depression, but modifications such as the growing emphasis on security appeared. The status system accompanying industrialization described in full detail in the first volume remained influential even in spite of the loss of funds for purchasing luxury goods.

Using this set of studies offers slightly different problems from those involved in identifying the generalized implications of Park's theory. Where Park failed to allow for patterns of urbanization by formulating his observations of Chicago subcommunities in general terms, the Lynds obscure the possibility of conceptualizing diverse patterns of industrialization by presenting their impression of the changes it wrought in Muncie in deceptively specific terms. Anyone wishing to trace the same processes studied by the Lynds in other communities would do well to observe closely differences arising when breakdown of the craft is slower or is resisted more vigorously than in Muncie. Other differences include the refusal to recognize commercial status symbols by traditional community leaders. This actually happened in Yankee City, where the upper-upper-class "old" families refused to accept the status claims of newly wealthy members of the lower-upper-class.

Students of industrialization in the sixties must attend to technological innovations like automation which promise to disrupt the job structure almost as much as the breakdown of the craft hierarchy. It is still too soon to determine the full effects of this novel phenomenon on community patterns, but anyone wishing to do so could profitably follow out the implications for status arrangements along the lines indicated by the Lynds. Here, too, the circumstances of contemporary mass marketing in which advertising, merchandising, and "planned obsolescence" play an even bigger part than in 1936 would have to be scrutinized closely to pick up the other phase of the industrialization cycle.

CHAPTER 3

LLOYD WARNER AND BUREAUCRATIZATION
IN YANKEE CITY

THE THIRD classic study of an American community, ranking with those of Chicago and Muncie, is Lloyd Warner's exploration of Newburyport, Massachusetts. This has been reported in four volumes of an announced six-volume series. The field work was completed between 1930 and 1935 by a research team with exceptional facilities. Of the four volumes published, *The Social System of the Modern Factory*, by Warner and J.O. Low, makes the most significant contribution to the theory being developed in this paper. Two other volumes, *The Social Life of a Modern Community*, by Warner and Paul Lunt, and *The Social Systems of American Ethnic Groups*, by Warner and Leo Srole, will be discussed briefly to provide background for a more detailed analysis of the fourth volume.

Warner's reasons for choosing Newburyport are summed up in the following passage: "We sought above all, a well-integrated community, where the various parts of the society were functioning with comparative ease. We did not want a city where the ordinary daily relations of the inhabitants were in confusion or in conflict. We desired a town whose population was predominantly old American, since our interest was in seeing how the stock which is usually thought of as the core of modern America—the group which ordinarily assimilates the newer ethnic groups—organizes its behavior when not suffering from an overpowering impact of other ethnic groups. A community was sought which had a long tradition, that is, where the social organization had become firmly organized and the relations of the various members of the society exactly placed and known by the individuals who made up the group. We wanted a group which had not undergone such rapid social change that the disruptive factors would be more important than those which maintained a balanced grouping of the members of the society."[1] Furthermore, he wanted some industry, relative autonomy, and a number of different ethnic groups. Like the Lynds, he too wanted a community small enough to be studied as a totality.

The above passage makes it clear that Warner hoped to eliminate from his study the extreme effects of urbanization. He did not want a community that had grown rapidly due to ethnic immigration but rather one where the relationships between various ethnic groups and the Old American core were established gradually over a long period of time. Furthermore, his desire for a settled community meant that the economy and life ways were to be disrupted as little as possible by industrialization. In short, he was interested mainly in social organization and hoped to eliminate disorganization. As it turned out, this was not possible, but his effort at doing so led him to a kind of community in which a third basic social process—bureaucratization—could be seen working together with industrialization and urbanization to transform the social structure of an American town.

The first volume, *The Social Life of a Modern Community*, offers a great deal more information about the inception of the study, the methods used, and the theoretical framework than do most other community studies. Warner's explicitness in these matters could well be emulated. However, the theoretical framework set forth is really the general orientation to social structure of a social anthropologist instead of a series of substantive hypotheses. In Chapter Five the author mentions starting with a set of such hypotheses, all of which apparently

[1] Warner, W. Lloyd, and Lunt, Paul S., *The Social Life of a Modern Community*, Yale University Press, New Haven, Conn., 1941, p. 38.

centered around the assumption that division into economic classes would satisfactorily identify structural groups in this New England community of 17,000. He observes that this early assumption was soon proved inadequate as disparities in social ranking turned up with surprising frequency. People with little wealth sometimes had social status which was high and vice versa. More interestingly, the field workers soon began to locate a number of ranked groups, more or less accurately perceived as such by many Yankee City dwellers and rather closely correlated with geographical portions of the city.

As elaborated in Chapter Six, this became the basis for conceptualization of the class structure of Yankee City as a six-class system. This is Warner's familiar division into upper and lower grouping respectively of each of three classes: the upper, middle, and lower. Much of the first volume is devoted to establishing the existence of these six groupings and showing their ramifications, in each instance, through several aspects of life style. There is no need for going into the details of Warner's highly ingenious techniques for locating people in the class structure, except to notice that in later formulations he combines "objective" measures like occupation and wealth with "subjective" factors like rank accorded by others and ranking of associational memberships. He calls the first, Index of Status Characteristics, and the second, Index of Evaluated Participation. Warner's efforts at establishing baselines in terms of these indices suitable for classifying the whole American population have been criticized so frequently that there is no reason to do so again. However, it will be taken for granted that this six-class system was useful and valid in Newburyport at the time of his research.

In relation to earlier studies, Warner's contribution remains the identification and systematic exploration of community status systems. The Lynds were well aware that the business class in Muncie had generally higher prestige and belonged to higher status organizations than the working class, but they did not develop conceptual tools for examining this as sharply as did Warner. The picture of social mobility in Newburyport is far more carefully drawn than in Muncie. In addition, the existence of a traditionally prestigeful upper-upper class adds an important dimension to the study of social stratification—that of regional variation in social structure. In general, one would expect to find these "old families" most common in New England and the South, but similar formations of this type do occasionally appear elsewhere.

From the standpoint of the theory being developed here, Warner's interpretation of community status systems as resting on a configuration that includes associational membership, life style, and occupation permits a deeper interpretation of social mobility and the conflicts it entails than simpler formulations. It provides a kind of socio-cultural context for understanding "why they work so hard" which cuts even more deeply into the motives of both workers and businessmen than did the Lynds' analysis.

Warner had been well prepared for his discovery of the link between status, life style, and associational membership by his anthropological training. Furthermore, his determination to study a community with its traditional system of social organization intact, insured the identification of such relationships. Perhaps the most interesting unanticipated finding of this study was not the discovery that economic position did not determine social position with any complete certainty, but rather his opening up the question of the relationships between economic position, social position, and, in later volumes, political power. Once the relationship between the three is seen to be more complicated than one of any simple one-way determination, then patterns of interrelationship can be extracted for study.

It must be admitted that Warner did not set out to do this explicitly in his first volume. Here, he took interrelationship for granted and interpreted the position of each class according to special criteria. The upper-upper, for example, held their

[73]

high status because they were descendants of the "founding families" who had always been the aristocrats of Yankee City. They were taught aristocratic manners from the cradle and inherited membership in the high-prestige associations of the town. Their homes had been in the family for generations, thereby becoming a prestige symbol that money literally could not buy. The lower-uppers usually were people who had acquired wealth at later points in their familial history so that they could not lay claim to honorable lineages as did the old families. One of the main points of interest in this first volume became the description of the efforts made by these "new rich" to acquire the prestige symbols that would identify them as members of the class just above them.

Since Warner does not describe the history of the community in any detail, the reader of the first volume has to reconstruct his own historical picture. He presents a cross-sectional description of the six classes, confining his dynamic comments to Chapter Seven, which consists of vignettes describing sequences of events in the lives of persons from each of the classes. These are arranged to illuminate aspects of their behavior which are especially linked to their social status. In particular, Warner deals with status problems, status defenses, and techniques of mobility. From the profiles the reader gets some sense of how it feels to be a member of each of the six classes.

Warner also gathered a great deal of statistical data on the people of Newburyport. Every member of the community had his social characteristics reported on cards which were processed so that the shared social characteristics of each class could be identified. Though there is not as much interpretation of these characteristics as would have been desirable, they do demonstrate the existence of the postulated set of six classes. Furthermore, they show how the ethnic groups, associations, churches, and other community institutions fit into this class structure.

In dealing with associational life, Warner treats it as a fundamental mechanism of integration in the community. All of the social categories in Yankee City have characteristic associations and these often become the main link between different categories. Warner differentiates several types of association according to the ages, sex, class, ethnic or religious affiliation of members, as well as substantive distinctions in terms of the institutional area in which the association operates. He has a complicated typology of associations which includes further differentiations according to scope, degree of closeness, autonomy, and the like. There are two points of special importance. First, some of these associations cut across class lines and thereby serve to facilitate mobility as well as communication between classes. Second, some of the associations link Yankee City to larger social aggregates providing regional and national ties for local interest groups.

Volume Two of the series, *The Status System of a Modern Community*, adds little of consequence for our purposes beyond reinforcing the conception of the way in which the six-class status network organizes the affairs of the community in each area of institutional life. Volume Three, *The Social Systems of American Ethnic Groups*, however, does add some elements worth noting. It provides a detailed picture of the assimilation process in this comparatively non-urban community which contrasts neatly with the picture of this same process in Chicago.

This third volume differs from the first two in that far greater historical depth is achieved. By focusing on the processes of assimilation and segregation in the context of a community where the social structure had previously been carefully elucidated, it became possible to delineate the relationship of each ethnic group to the whole community. Adding a historical dimension helped to show how this relationship changed as the pressure to Americanize took hold and exposed the various institutional mechanisms whereby this pressure was resisted.

[75]

I. FOUNDATIONS

As in the first volume, one of the most valuable chapters is the set of profiles describing the details of ethnic life. Here the authors present five vignettes portraying incidents from the lives of Irish Catholics in Newburyport in such a fashion that they illuminate the status conflicts and adjustments of persons with this background from each of the six classes. It is clear that the social system of lower-class Irish Catholics differs from that of those who have attained middle- or upper-class status. The vignettes suggest something of the complexity of these ethnic social systems when their interrelationship with the status problems of the various classes is taken into account.

Warner found that it was possible to locate the general position of each of the eight ethnic groups in the whole structure even though members of each could be found at different places. He distinguishes between the mobility of the whole group and the mobility of individual members of each group and examines the interrelations between the two. The groups are compared in terms of the steps and processes according to which they have:

"a. progressively advanced in the major status hierarchies of Yankee City and

"b. progressively adapted the internal organization of their community systems."[2]

Progress made by the various groups in terms of access to desirable residential areas, occupations, and social-class attributes is described in detail. The summary of the last set of findings is worth quoting in detail:

"To summarize, the degree of ethnic approximation to the statuses of the natives is correlated primarily with the length of the group's establishment in Yankee City. That is, all groups have progressively climbed toward higher positions in the three hierarchies. However, certain secondary factors have produced differences in the rates of mobility among the various groups:

[2] Warner, W. Lloyd, and Srole, Leo, *The Social Systems of American Ethnic Groups*, Yale University Press, New Haven, Conn., 1945, p. 32.

"A. Factors for retardation of status mobility

"1. Original migrational intention of temporary settlement (South Italians, Greeks, Poles).

"2. Family structure with patterns of maintaining customary status and of parental determination of status (French Canadians).

"3. Order of a group's appearance in the city, both because the earliest group encounters local conditions which no longer operate when later groups arrive and because, to a certain extent, the earliest group reduced resistance to and cleared the way for the advance of later groups (Irish).

"4. Large group population, a condition increasing the resistance to mobility (Irish and French Canadians).

"5. Proximity to the homeland, a factor for the slowing of the acculturative processes and therefore for the curbing of status advance (French Canadians).

"B. Factors for acceleration of status mobility

"1. Similarities between the ethnic ancestral society and Yankee City in general social-organization type (Jews).

"2. Similarities between the ethnic ancestral society and Yankee City in the religious aspect of culture (Armenians)."[3]

The detailed discussion of these various factors shows how they operate in the fashion indicated in the context of Yankee City and suggests that they would behave similarly in most community contexts. When ethnic groups have distributed themselves through several class positions, Warner notes that they tend to identify more closely with their class than with fellow ethnics.

Warner's analysis is always oriented toward differences between the several generations so that each group is shown to have two or more systems, with that of the younger generation always being more assimilated to the American ideal than the older. Each specific institution in the ethnic system is examined and its impact on the generations studied, from which

[3] *Ibid.*, pp. 101-102.

it becomes clear that the pressure toward "becoming American" in Yankee City rests upon the adoption of American consumption patterns of the same type that Lynd saw pervading Middletown life. The motive for abandoning old world life styles is the same as the incentive that pulled Middletown workers out of their craft roles: the desire for fashionable mass-produced commodities in ever-increasing quantities.

Again, the role of associations is highlighted and here the shift from "closed" ethnic associations to "open" American groups is seen to be central. Only the church is more significant as a basis for defending the integrity of ethnic patterns. The subtle processes whereby ethnic churches begin to lose their distinctive features while exclusive ethnic associations become increasingly confined to the immigrant generation are analyzed effectively in terms of the relations between sub-community and the larger community. This relationship is summarized:

"The forces which are most potent both in forming and changing the ethnic groups emanate from the institutions of the dominant American social system. Our political organization permits all adults to be equal within its structure. Although at first this equality is largely theoretical, it gives the ethnic members an attainable goal as the political success of the Irish, Germans, Scandinavians, and Italians demonstrates. Our developing industrial and factory economy with its own hierarchy permits and demands that ethnic members move up and out of their ethnic subsystems into the common life of America. The public school teaches the people to adjust to the central core of our life, provides them with technical skills for their own advancement, and gives them some of the power necessary to become upward mobile in our class order. The school, in belief and partly in practice, expresses the basic principles of American democracy where all men are equal; when the school cannot make them equal it struggles to make them culturally alike.

"The American family system breaks down and builds up ethnic subsystems. The ethnic parent tries to orient the child to an ethnic past, but the child often insists on being more American than Americans. Marriage also may maintain or disrupt the ethnic way of life. At marriage an individual may move out of his ethnic group into that of his spouse; or an individual who has become partly American may re-identify with his ethnic group and become more ethnic than in the past.

"Cliques and associations also operate to increase or decrease ethnic identification. If the child in school becomes a part of an American clique he is likely to move rather rapidly into the American way of life. On the other hand, if he is rejected and forced to participate in ethnic cliques he may become closely identified with the cultural group of his parent. This is also true for adult cliques and for adult associations. On the whole, however, the evidence from Yankee City shows that cliques and associations increase the participation of ethnic people in the life of the larger community and accelerate assimilation.

"Our class system functions for a large proportion of ethnics to destroy the ethnic subsystems and to increase assimilation. The mobile ethnic is much more likely to be assimilated than the non-mobile one. The latter retains many of the social characteristics of his homeland. Most ethnics are in lower social levels. Some of them become self-sufficient, interact among themselves, and thereby reinforce their old ways of life. Some of the unsuccessfully mobile turn hostile to the host culture, develop increased feelings of loyalty to their ethnic traditions, become active in maintaining their ethnic subsystems, and prevent others from becoming assimilated. But, generally speaking, our class order disunites ethnic groups and accelerates their assimilation."[4]

This all too brief summary includes many of the more important generalizations about the integration of various gen-

[4] Ibid., pp. 283-284.

[79]

erations of ethnic groups into Yankee City life. It is important to notice, however, that there are alternative outcomes, some yielding greater integration and others greater separation. We still need to know a good deal more about the conditions that in various community contexts lead to one alternative or the other.

The most important volume of this series is the fourth, *The Social System of the Modern Factory*. Warner opens this book with a chapter announcing the specific topic with which it is to deal, namely, "The Strike—Why Did It Happen." Here the focus of the community study is sharpening, so that a particular crisis in the life of the community becomes the subject of analysis. In contrast to earlier emphasis on social organization, Warner now announces:

"The best of all possible moments to achieve insight into the life of a human being is during a fundamental crisis when he is faced with grave decisions which can mean ruin and despair or success and happiness for him. In such crises men reveal what they are and often betray their innermost secrets in a way they never do and never can when life moves placidly and easily. If this is true for the study of men as individuals, it applies even more forcefully to the study of men in groups. It is when hell breaks loose and all men do their worst and best that the powerful forces which organize and control human society are revealed."

". . . We have selected such a crisis in the relations of management and workers in a factory of Yankee City for analysis. The study of this dramatic conflict illuminates the normal position of the factory in the community and the relations of management with labor."[5]

This assumption that the underlying or latent structure of a community will be especially visible in moments of crisis is a powerful analytic tool, as is indicated by the fact that here

[5] Warner, W. Lloyd, and Low, J.O., *The Social System of the Modern Factory*, Yale University Press, New Haven, Conn., 1947, p. 1.

Warner probes more deeply into the dynamic forces at work in Yankee City than he was able to in the earlier volumes.

Warner approaches the strike entirely as a community-wide occurrence, and each of the questions he raises about it has the community as a major point of reference:

"1. In a community where there had been very few strikes and no successful ones, why did the workers in all of the factories of the largest industry of the community strike, win all their demands and, after a severe struggle, soundly defeat management?

"2. In a community where unions had previously tried and failed to gain more than a weak foothold and where there had never been a strong union, why was a union successful in separating the workers from management?

"3. Why was the union successful in maintaining the organization despite the intense and prolonged efforts of management to prevent unionization and to halt the continuation of the shoe union?

"4. Why did Yankee City change from a non-union to a union town?"[6]

Through these questions, Warner transforms what could have remained a simple study of a strike into an analysis of basic patterns of community structure. He interprets the strike and acceptance of the union as symbols of the fundamental changes in relationships between two major groups, management and labor—changes that are then linked to the breakdown of traditional patterns in Yankee City.

It is worth noting that this volume is the first one in which the authors introduce the depression as a force affecting community life. Furthermore, this volume is the first to adopt a historical perspective comparable to the Lynds'. While the volume on ethnic groups did present a historical model for examining assimilation, it did not satisfactorily place this model within the context of the changing community. Volume Four

[6] *Ibid.*, pp. 6-7.

contains a picture of the changing social structure of Yankee City from 1800 onward, with an emphasis on the same period (1880-1935) covered by the Middletown studies. Warner does not link the depression explicitly to the more fundamental changes, but his data is sufficient to do so easily.

Warner treats the impact of the depression on the work situation, showing how the crisis laid the groundwork for the strike. Using vignettes, Warner describes the workers' growing hostility toward the upper classes who continue to lead secure lives while their own jobs are being so profoundly threatened. Some Yankee City workers actually saw the depression as a deliberate conspiracy by managers and owners against them. Clearly emotional reactions against the upper classes were here far stronger than any reported in Muncie and the specific feeling of being "exploited" far more prevalent.

At no point in this volume does Warner minimize the role played by economic factors. He reports the workers' grievances, mainly formulated in economic terms, and recounts the history of the strike, showing how the union seized on these grievances and coordinated the strike so as to pursue it to a successful conclusion: "Plainly, labor had won its first strike in Yankee City, and, even more plainly, an industrial union had invaded the city for the first time and had become the recognized champion of the workers. When searching for the answers to why such significant, new changes could occur in Yankee City, the evidence is clear that economic factors are of prime importance. But before we are content to accept them as the only answers to our problems, let us once more remind ourselves (1) that there had been severe depressions and low wages before and the union had failed to organize the workers, and (2) that the last and most powerful strike which preceded the present one occurred not in a depression but during a boom period when wages were high and economic conditions were excellent. Other factors are necessary and must be found if we are to understand the strike and have a full explanation of why it occurred and took the course that it did."[7] These are the clues that sent him deeper into the changing community than a less thorough analyst might have been impelled to delve.

The search for additional causes begins with a chapter on the over-all socio-economic development of Yankee City. Warner describes this development succinctly:

"The changes in the form of division of labor (see Chart 1) are another story of the utmost importance. In the beginning, the family made its own shoes, or a high-skilled artisan, the cobbler, made shoes for the family. In time, several families divided the high-skilled jobs among themselves, and later one man assigned the skilled jobs to a few men and their families. Ultimately, a central factory developed and the jobs were divided into a large number of systematized low-skilled jobs. The history of ownership and control is correlated with the changes in the division of labor. In early days, tools, skills, and materials were possessed by the family; eventually, the materials were supplied by the owner-manager, and soon he also owned the tools and machines. The sequence of development of producer-consumer relations tells a similar story. The family produced and consumed its shoes all within the circle of its simple unit. Then, the local community was the consumer-producer unit, and ultimately the market became national and even worldwide. Worker relations (see Chart 1) changed from those of kinship and family ties to those of occupation where apprenticeship and craftsmanship relations were superseded and the individual unit became dominant in organizing the affairs of the workers. The structure of economic relations changed from the immediate family into a local hierarchy and the locally owned factory into a vast, complex system owned, managed and dominated by New York City."[8]

Out of this historical complex, certain central elements can be identified. From the standpoint of this theory, it is the

[7] *Ibid.*, p. 53.   [8] *Ibid.*, p. 64.

growth of absentee ownership which proves most significant.

Like the Lynds in *Middletown*, Warner grasped the central importance of the breakdown of the craft hierarchy, placing this factor first in his analysis of non-economic causes. Most jobs in the Yankee City shoe factories required little or no skill after mechanization and the introduction of assembly-line technique. As in the glass factory, workers no longer could rise through gradual acquisition of skills to positions of maximum skill and prestige. With mechanization, the foremen became personnel manipulators instead of senior craftsmen. This led to their being selected on the basis of administrative capacities gained in schools rather than on the job. The relation between age and social position existing in a craft community, whereby the older workers have more prestige, authority, and security, was disrupted. Thus the technological changes in production, mechanization, and assembly-line specialization associated with industrialization had an impact on Newburyport similar to that experienced in Muncie. Warner does not link this to the impact of industrialized consumption patterns, but the data in Volume One could be used to clarify this somewhat. Certainly even the lower and middle classes are dependent on mass-produced commodities to maintain their respective styles of life and to move toward the life style of the next highest class.

Warner's analysis of the character of mechanized jobs is a bit more emphatically concerned with the effects on social organization than the Lynds'. He directs attention to the problems of social control in the shop so that industrialization is seen as disrupting the old craft control systems, on the one hand, and encouraging bureaucratic controls, on the other. He analyzes the effects of ethnic solidarity and shows how management uses the insecure community status of the newer ethnic groups to render them more compliant. Other bases for worker solidarity, such as skill differentials and wage differentials, are analyzed with special reference to the specialties in which management evaluations differ from those of the workers and the community. There proves to be no consistent relation between wages, prestige accorded a job by the workers, and the fact of it being a hand or machine job.

Taking off from his interest in social control and his determination to interpret changes in the factory in terms of the changing community, Warner presents the decisive variables in his analysis of the strike in Chapters Seven to Nine. Seven shows how Yankee City factories have fallen into the hands of absentee owners. Eight takes up the differences between the old managers and owners and the new absentee owners. And Nine deals with the loss of status experienced by the shoe workers after industrialization and bureaucratization in such a fashion as to indicate how the latter especially was a key causal variable in the 1933 strike.

Like many other industries after the turn of the century, Yankee City shoe factories were caught up in the process of centralization of ownership. The conditions of production for a mass market were such as to require the capital and distributive facilities of large-scale enterprise. Gradually the locally owned factories were being absorbed into large chains. By the time of the strike, most of them were controlled by New York city financiers and run by "outsider" managers. These facts of American economic life are quite familiar, but Warner's distinctive contribution lies in showing how they alter the social structure of the local community.

One obvious consequence, the not-so-obvious further implications of which Warner spells out in some detail, is that the community's control over factory decision-making is considerably weakened. Managers of local plants are not part of local prestige structures and they look to extra-local agencies, especially the higher executives of their corporations, for confirmation of their status claims. To be successful, they eventually have to leave local plants for the higher reaches of the corporate hierarchy. Their interest in the community remains primarily focused on the extent to which it provides a tractable labor force. Basic policy decisions, some of which can drastically affect community affairs, are made at top echelons, leaving local managers to enforce unpopular policies without giving them much discretion in determining their character. The profit-making logic determines most decisions, yet this does not necessarily coincide with the sentiments or interests of persons at lower positions in the hierarchy. The function of the bureaucratic hierarchy is to guarantee that top-level decisions will be effectively carried out despite conflicting sentiments at lower levels. Actually both workers and managers at lower levels are not supposed to have the kind of discretion or responsibility that would allow them to interfere with such decisions.

Local concerns are affected by large-scale enterprises when they join horizontally organized manufacturers' associations through which the firms in a single industry within a region or the entire nation join in making over-all policy decisions. This assumes that all shoe manufacturers have certain interests in common and can benefit from mutual association. Warner shows that this model of organization parallels that of the industrial union, and suggests that the logic of this kind of unionism as against the distinctions maintained by craft unions involves denying the sentiments on which local class structure is based. It admits all shoe workers regardless of class or ethnic position and affirms their common interests as against those of management. In so doing, Yankee City workers declare their solidarity with shoe workers in other towns and cities.

Acceptance of such extra-local affiliations by managers or workers in the Yankee City shoe industry would have been unthinkable during an earlier era. It is this earlier era that Warner deals with in his historical sections. The business leaders of the community, Caleb Choate, Godfrey Weatherby, and William Pierce, were all men from prominent local families whose status as creative entrepreneurs went unchallenged. They were supposed to have held the interests of both their employees and Yankee City above immediate economic gain, and their deep involvement in every aspect of community life suggests that at least the latter sentiment was probably true. During the strike there was frequent comment by the workers that such action would not have been necessary if these traditional figures had lived and retained control. They became heroes with whom the managers and owners of the thirties could not hope to compare. During the strike anecdotes about the close personal relations between these dead community leaders and their employees were widely circulated.

One fact emerging clearly from Warner's data is that the absentee owners were largely Jewish and therefore had this stigma added to their status as "big-city outsiders." The symbolic juxtaposition of the Bronsteins and the Lutskis against the Choates and Weatherbys of an earlier era was clearly important in mobilizing the hostilities of the workers. The managers, even though a few had local ties, were unable to rectify matters. Those who were outsiders had almost the same stigmas as the absentee owners and, besides, their power to negotiate was limited by their intermediate status. The few managers who had local ties were also lacking in decisive power. In addition, they had not earned their authority by creative entrepreneurship, as had their predecessors, but rather appeared as lackeys of alien forces.

Thus, the strike in Yankee City is seen as the result of a transition from Local to Big City controlled capital and acceptance of the union as spokesman for the workers as part of the same process. It entailed the disruption of a traditionally organized community in which factory authority was reinforced by leadership in all areas of life. Choate, Weatherby, and Pierce were also subject to local control: ". . . first, they were dominated by local sentiments which motivated them 'to take care of their own people'; second, they were under the powerful influence of the numerous organizations to which they belonged; and, third, their personal contacts with the local citizens directly related them to influences from every part of the city."[9]

The descendants of these men, reduced to the status of managers or powerless stockholders, could not exert nearly the same influence over the workers, even though they retained some general prestige in the community. Warner suggests that a few of them actually wanted to side with the workers against the aliens. An integrated social system wherein high prestige had been correlated with high economic status and high influence had been disrupted by the increase in centralization affecting the town's main industry. Established patterns of interrelationship between workers and owner-managers, between low and high status townspeople, had been similarly affected: "The longing for the idealized past when men had self-respect and security was symbolized in the three dead owners; and these symbols materially aided the workers in defeating management since the workers and management felt that the present men could not match the 'gods' of the past. The workers and managers in the shoe industry had lost their sense of worth and mutual loyalty. No longer were they men who had a common way of life in which each did what he had to do, and, in so doing, worked for himself and for the well-being of all."[10]

This interpretation of the strike locates it in the framework of larger social changes of a kind usually studied under the topic of bureaucratization. Lynd picked up one aspect of the problem when he speculated as to why men should work at jobs that provided little intrinsic satisfaction. Shifting the focus to social organization, Warner asks why they should work for men whose authority they do not accept.

Both Warner and Lynd recognize that workers lost status as well as opportunities for mobility with the breakdown of the craft system. Warner notes the possibility that unions can compensate for this somewhat. There is always the possibility of election or appointment to union office. Furthermore, unions

[9] *Ibid.*, p. 152.   [10] *Ibid.*, pp. 157-158.

lift the status of workers as a group by giving them some kind of representation in dealing with the high-level owner-management groups in terms to which these groups must pay attention. Instead of the individual worker rising through the ranks, you have a kind of collective mobility whereby all the workers in an industry raise their standards of living by bargaining for wage increases with an equally centralized association of managers. The union has its highest echelons as far removed from Yankee City as those of management, but these union top echelons must be more responsive to local pressures because they depend on them for support. The union is essentially a responsive counter-bureaucracy through which workers as a group improve their common lot.

In the following passage, Warner shows how the workers' anxieties about their status arose from their changed situation and shaped their participation in the strike: "The workers of Yankee City were able to strike, maintain their solidarity, and in a sense flee to the protection of the unions because the disappearance of craftsmanship and the decreasing opportunities for social mobility had made them more alike, with common problems and common hostilities against management. The craft differences had been wiped out, and occupation mobility in the craft hierarchy and, secondarily, social mobility in the community had been stopped. The workers felt even more alike and were increasingly motivated to act together because their new occupational status had contributed to their downward orientation in the community. There is no doubt that each worker's uneasiness and reasoning about what was happening to the status of his family and himself—a situation which he meagerly comprehended and which was almost beyond his ability to communicate coherently to his fellows— whipped him into attacking the owners, who provided visible targets and could be held responsible for the loss and degradation of the worker's cherished way of life."[11]

[11] *Ibid.*, pp. 171-172.

[89]

It was the sense of downward mobility over a generation that provided the background for the willingness to strike and organize. The presence of suitable symbols in the form of the "godlike" old owner-manager who contrasted so sharply with the "little men and aliens" who had taken over added to the extreme provocation furnished by the depression, and the ready availability of the union provided the matrix out of which the strike grew. There is good reason to assume that the workers exaggerated the virtues of the old "gods" just as they overstated the vices of the new owners. And they also probably magnified the degree of solidarity and communality across class lines that existed in the "good old days." But this is certainly not the first time that half-truths about "Golden Ages" or "Dead Heroes" played a role in history.

Warner's awareness of the problems involved in legitimating managerial authority is outstanding. Perhaps because he approached the factory in terms of changes in the community, he was able to see that the breakdown of traditional systems of reciprocal expectation such as had obtained between management and labor in the early Yankee City factories did not ensure the immediate acceptance of absentee managerial prerogatives. Instead, these prerogatives and the logic of profitmaking that they assumed were resisted vigorously to the point where the shoe workers were willing to align themselves with another extra-local force—the union—against the "outside" managers. Relations between the two groups in Yankee City were thenceforth to be mediated by impersonal contracts arrived at through negotiation between representatives of the top echelons of the two bureaucracies—management and labor. Relations between the local members of the two groups were drastically revised, reflecting the way both lost the power for decisive action. The old system of social organization in which local economic, political, and social positions were in regular alignment had broken down as people in all positions of the local hierarchy found themselves dependent on extra-local influences. Simultaneously, their sense of belonging to a common enterprise, whether it be the shoe factory or the city itself, seriously declined. As impersonal mechanisms are instituted to regulate their relationships in the factory, so do similar mechanisms appear to keep them segregated in the community. One might expect a decrease in associations cutting sharply across class lines.

It is logical to suppose that the reactions to invasion by absentee owners and the introduction of bureaucratic coordination in Yankee City were especially aggravated by the persistence of the traditional system of social stratification summarized in Warner's distinction between upper-uppers and lower-uppers. Thus, the people of Muncie never reacted as strongly to absentee control in their GM plant, perhaps because the automobile industry had always operated on this basis. There was no lingering craft tradition and the work process had always proceeded along impersonal bureaucratized lines. Thus, the extent to which this process is felt as a trauma by a community depends on the prior organization of the work process. It is doubtful if there are too many communities which have experienced the close integration at the leadership level of the three hierarchies of power, economic wealth, and prestige that lay in the near past in Yankee City. Even the X family in Muncie did not have traditional claims to high position.

On the basis of the dynamic patterns identified in this volume, one cannot help but wonder if the system of social stratification presented in the first volume will not really turn out to be a transitional phenomenon. Just as there is reason to believe that the concentration of power in the hands of the X family in Muncie did not continue long after the depression, so it is likely that the prestige of the old families of Newburyport must have waned somewhat with the rise to economic power of the then emerging lower-uppers. Muncie probably now has more absentee power than in 1936, an absentee power integrated with the X's mechanisms of domination, while Newburyport absentee managers are probably beginning to over-shadow the declining "aristocracy." Certainly the vignettes in the first volume suggest that the upper-upper class prestige monopoly was under attack and their failure to provide leader-ship during the strike must have weakened it even further. It is doubtful if the six-class system found in the thirties would be as applicable to Newburyport in the fifties. Yet it was just this exceptional degree of traditional social organization that made possible the study of the more extreme effects of bureauc-ratization.

From the several volumes of the Yankee City series, we get a picture of the complex social changes occurring when a com-munity with a traditional "Old American" social structure is taken over by absentee owners and managers. In particular, the relations between workers and managers are analyzed so as to elucidate the concomitants of the shift from traditional to bureaucratic patterns of authority. The old feeling of soli-darity based on a sense that everyone in town belong to a common community gives way to sub-communities with hos-tile attitudes toward each other. This division along class lines is symbolized by the rise of the union; however, the union also helps to stabilize relations between the classes along rational-bureaucratic lines.

Yankee City was also affected by urbanization and indus-trialization. In Volume Four we get an excellent opportunity to examine the relations between these two environmental pres-sures and the dominant pressure in this community—bureauc-ratization. Warner shows that the conjunction between assem-bly-line technology with its concomitant status challenges to the social status of craft workers and the impersonal authority system which this industrial technology brings with it serves to "alienate" the worker and leave him prone to affiliate with trade unions as a vehicle for counter-assertion. But his relation to the union remains almost as impersonal as his relation to the absentee-controlled factory, and he does not regain any

[92]

real sense of participation in a joint enterprise such as the old shoe craftsmen were able to enjoy.

Just as the workers find themselves being forced to accept absentee domination and separation from the community as a whole, so are the "old families" being replaced by the newly-wealthy ethnic merchants and the newly-arrived managers as community leaders. Traditional status claims lose their significance as the social structure of the town moves into line with national patterns, according to which wealth conspicuously displayed is the main avenue to status. The old families lose their economic base at the same time that their political influence declines and, accordingly, their prestige necessarily drops. Warner allows us to get considerable insight into the interrelationship of these changes.

Apparently, urbanization in Newburyport did not create much disorganization, since mechanisms for assimilating the incoming ethnic groups were not strained by the rate of immigration. Different ethnic worlds appeared and began to transform themselves as the demands of mobility required, but the opportunities for doing so without interference were sufficient to prevent widespread disorganization. Without tremendous or unusually rapid increases in size, ethnic diversification and the beginnings of assimilation appear. Even more important, urban impersonality takes hold and Warner's data certainly suggest that bureaucratization with centralized absentee control over community affairs is an important step in this process. The classes in Yankee City—once joint participants in a common communal system—now confront each other as embodiments of collectivities under the control of remote power centers which determine their relationships to each other far more than factors arising in local life. The depression furthers this feeling of local impotence and shows how a historical crisis reinforces the interlocking structural consequences of urbanization, industrialization, and bureaucratization in this New England town.

[93]

TOWARD A THEORY
OF AMERICAN COMMUNITIES

AFTER these selective summaries of three sets of American community studies, the remaining task is to show how they provide the basis for developing a generalized theoretical approach to the field. This theory should be able to account for the main patterns of change found in each of these communities so as to facilitate exploration of the same patterns under quite different circumstances. In other words, the theory has to point toward further cumulative research. It is just this problem of accumulation that has made theorizing in this area so difficult. Community studies are always confined in time and space so that the relevance of any particular investigation for understanding later or earlier studies is always in question.

Even if one assumes that time can be held in abeyance, it is doubtful that the student of any particular American community can reasonably claim to generalize his distinctively sociological findings so as to imply descriptive validity for all or even most American communities. The search for "representative" communities on the basis of which generalization about all of them or about the society itself can be developed is doomed to failure. Yet Park, the Lynds, and Warner all made this assumption about the objects of their research. The point is not that Chicago, Muncie, and Newburyport were representative communities in any statistical sense but rather that they were undergoing processes of structural transformation that affected all American cities and towns to one or another degree, and therefore could be used as laboratories in which to study these representative social processes.

[94]

The emphasis in the foregoing chapters on one process in each community—urbanization in Chicago, industrialization in Muncie, and bureaucratization in Newburyport—suggests the direction in which this effort at community theorizing will move. While Park was explicit about searching for generalizations about urbanization and Warner was aware of the general implications of his research (though he never explicitly placed it in the framework of a theory of bureaucracy), the Lynds confined themselves more closely to the enormous task of reporting what happened when industrialization hit Middletown. They left to the reader or later theorist the further task of deciding which of the structural changes were likely to accompany industrialization when it occurred in other places.

Unfortunately, sociological readers of the Lynd study have simply not been diligent enough. *Middletown* is now read mainly by undergraduates as a sociological classic of the kind everyone should be familiar with even though its significance is primarily historical. After all, the Lynds describe events that took place between thirty and sixty-five years ago and the real interests of undergraduates, no less than of professional sociologists, are usually in more recent matters. American sociology, unlike its European counterpart, has always displayed impatience with the ineradicable historicity of much sociological data. Why should anyone today be concerned with what happened in an obscure town like Muncie, Indiana, such a long time ago? Muncie itself has undoubtedly changed considerably since then, as the Lynds themselves demonstrated by returning in 1936 to produce *Middletown in Transition*, a book superseding the first volume—or does it?

The question conceals genuine problems. How can generalizations be extracted from material which is probably not representative and which refers to events that occurred long ago? Furthermore, of what use would these generalizations be for understanding that same community now, thirty years later, no less some other community as it was at that time or as it is today? It is safe to assume that even at that time the effects of industrialization on Chicago, with its complex ethnic and class structure, must have been significantly different from the effects of this same process on Muncie.

The problem of generalizing from the community study has long been a matter of concern to sociologists, and one suspects that the apparent failure to resolve it satisfactorily helps to account for the relative neglect of this field during the past fifteen years. Why study communities when one can much more easily study small groups? In spite of the recent proliferation of mechanical gadgets for recording interaction patterns, small groups are still far easier to observe systematically than small communities.

Furthermore, students of small groups can sidestep the problem of generalizability simply by defining their objects of study on a sufficiently high level of abstraction as to eliminate substantive differences of social context from the range of legitimate concerns. Presumably, a small group of miners in a mine can be analytically equated with a small group of soldiers on a battlefield and a small group of students solving a chess problem on the wrong side of a one-way screen. If the researcher is sufficiently ingenious, he can even arrange to examine the formation and dissolution of small groups at his own behest.

The advantages and perhaps the attractiveness of small-group research would seem to lie largely in the readiness with which observational studies can be conducted and findings generalized. In both of these respects, the field of community studies is not nearly as convenient. At its best, however, it offers closeness to human experience and human problems of a kind rarely captured in the confines of the small-groups laboratory. Even the literature appeals to one's appetite for diversity so that the susceptible reader is easily enthralled by the descriptions of the strange social worlds—the slums, the ghetto, the hobo jungles, Bohemia, the Gold Coast, and the taxi-dance halls—

that Park discovered during his explorations of Chicago. After sufficient immersion in the Lynds' *Middletown*, the people, homes, schools, and churches of Muncie acquire an elusive familiarity that renders them clearer than the persisting images of communities in which one may have spent many years.

There is no denying the sheer descriptive appeal of the community study. Susceptibility to this appeal, some feeling for the varieties of styles of living found in complex modern settings, is undoubtedly a real asset for students in this field. Such sensitivity, however, must be balanced by sustained concern with the general implications of this rich observational detail. William Foote Whyte, in *Street Corner Society*, presents as vivid a picture of slum dwellers and slum life as has ever been captured by any journalist or novelist, but the distinctively sociological merit of his work lies in the interpretation of these life patterns in terms of hypotheses about slum social structure, ethnic assimilation, and mobility. Novelists, journalists, and community sociologists share an interest in describing the life of their times, but we have a special responsibility based on our commitment to developing a reliable body of knowledge.

While writing the three highly abbreviated analyses of the Chicago studies, the Lynds' *Middletown* volumes, and the Yankee City Series, we could not help but be continually aware of exactly how much of the raw material was being omitted. In our effort to develop a somewhat more general theory, specific emphases in each of the sets of studies had to be extracted and conceptualized differently. This necessarily meant leaving much of the data untouched and even "distorting" some of that which was extracted when conceptualizing it differently than did the original field worker. Yet this is necessary to expose the relationship between the three sets of studies as well as their collective relationship to later research. Such generalization does not permanently damage the original researches, since these are always available to all who would study them. Hopefully, it should make such examination of the sources

[97]

more fruitful by placing them in a more general context and thereby illuminating the particulars reported in each.

Theorizing from such a vast amount of empirical material as is represented by the three sets of studies already reviewed, as well as from the entire body of American community studies, is a challenging and even frightening task. The person who would do so has to be prepared to omit much that seemed important to the authors of the studies he is working from as well as much that seems important in them to him but which does not yet fit into his evolving theoretical framework. Finding an adequate level of generality is an ever-present problem, as is remaining satisfied with any given level once it has proved useful. This theory attempts to stay as close to the empirical materials as it can while still establishing the relationship of the trends they reveal to the major social processes of our time. Thus, the choice of urbanization, industrialization, and bureaucratization as processes shaping American community life was based on the findings of general sociological theorists like Marx, Weber, Durkheim, and others. That each of these three sets of studies exemplifies one of the processes as they manifest themselves at the level of the community suggests that the field workers were able to detect major forces of change even though they did not always interpret them with these concepts.

The theory to be developed will block out an approach to *American* community development. Whether the same interpretations could be applied to community development in other national or cultural contexts will not be explored here. It is admittedly an historically limited theory. For that matter, one of its main requirements is that the historical context of any community study be carefully examined as a shaping force.

The problem of generalizing from community studies having a specific locus in time and space is to be met at least in part by explicit incorporation of the relevant historical events as structural influences. The Lynds' *Middletown in Transition* is an excellent example of a study that directly meets this requirement by building the research around the impact of a major historical crisis. Warner never succeeds in bringing the depression explicitly into his interpretations until the fourth volume, and even there he does not consider all of its important structural effects. In any event, the kind of theory being developed here emphasizes the "decade experiences" so that any American community study during the early forties would have to assess the impact of World War II, while one completed in the early fifties would be committed to including as one variable the effects of the cold war and, finally, the Korean War. The national and international events of the decade must be studied as they impinge on the community and the underlying forces shaping its structure.

This means that every community study is to be viewed as a case study. We can go even further and maintain that they should all be studies of the effects of basic processes and historical events on changing social patterns. This means that the state of affairs before the change as well as while it is in progress should be carefully specified. Every good community study is a study of transitional processes.

Here, probably, we come upor a main weakness of the Chicago studies. They presuppose a conception of rural social organization which gives way under the pressures toward urbanization without ever explicitly describing the earlier social system. The notion of a folk or rural-urban continuum attempts to do so on a generalized plane and does make some important theoretical contributions. Actually, the Chicago sociologists took rural social organization more effectively into account when they studied the adjustment problems of immigrants. And there are undoubtedly sound historical reasons why the reconstruction of Chicago's transition from a rural to an urban center was not brought to their attention systematically. The main shift had occurred well back in the nineteenth century and they were preoccupied with the problems of twentieth-century urbanism. It is important, however, that further studies of urbanization in its present form always go into sufficient depth to indicate the base-points of rural and intermediate urban social organization. In other words, the research and theorizing should always be formulated in terms of typical stages of development. Perhaps this is the same as Park's conception of Natural History.

It is also worth noting that this model presumes that the theoretical accumulation will pivot around growing insight into the processes of change in various contexts. Instead of searching for representative cases, the query will always be, "What does the case under scrutiny represent?" At the present state of our knowledge, we cannot legitimately assume knowledge of the social structure of any community before beginning research. The field worker has tools for identifying sub-communities and analyzing their relationships to each other. Furthermore, he has considerable knowledge about the bases for sub-community formation and some sense of the likely configurations. Even with all of this, it is necessary to recognize that each new community studied will represent a slightly different configuration. No one is likely to discover anything like a duplication of Muncie in the twenties or thirties today, nor can the social structure of Newburyport in the thirties be closely duplicated. Our set of structural models can be used to provide clues as to significant groupings but not as a handbook of the range of possibilities. It will take a great deal more research before such a comprehensive handbook can be made available and one can be certain that any such handbook will always be liable to obsolescence as novel patterns emerge. The interpretive focus has to be on the changing configuration and the social processes underlying the observed changes.

This kind of analysis has been called "field-theoretical" by Kurt Lewin and it involves a number of methodological assumptions that need not be taken up here. The main point is the recognition that every community can be viewed as an organized system standing in a determinate relationship

[100]

to its environment. The system-environment model then focuses attention on this relationship between the two. If the research is directed at changing patterns, then the sources of systemic disorganization, whether internal or environmental, must be carefully examined alóng with the mechanisms of reorganization.

Adequate analysis involves careful specification of the state of the system preceding the changes as well as during the change. This must always be accompanied by equally careful specification of the environmental pressures in the early state and their changes as they relate to changes in the system. Both sets of variables have to be dealt with as they interact to produce the changes and react to them.

To illustrate this, a brief reference to Elin Anderson's *We Americans* might help. This was a study of Burlington, Vermont, published in 1937 in which the author wanted to observe patterns of ethnic assimilation and segregation. For our purposes it can be viewed as a kind of case study of urbanization. In 1934 Burlington had a total population of 15,000, which placed it among the smaller American cities and made it therefore hardly comparable to Chicago in terms of total population size. Furthermore, the rate of increase from about 1880 onward had been fairly gradual, with approximately the same small numbers arriving every decade. Yet it did correspond to the Chicago pattern in one respect: the immigrants had come from a wide variety of ethnic backgrounds, resulting in a number of ghettos still maintained in 1937. Nonetheless, there was little apparent disorganization of the kind found in Chicago slums. If anything, Burlington appeared to be overorganized. Each ethnic community formed a tight circle and its members were bound by their ties to it as well as by the reluctance of the larger community to identify them in any other fashion. Without going into any further detail, it is clear that an analysis of urbanization in Burlington would have to study the mechanisms whereby the various ethnic groups were incorporated into the life of the town without being driven to dissipate their distinctive life styles, as was the case in Chicago and Newburyport. Hypotheses would have to be developed that referred to changes in the social structure of Burlington as it absorbed these immigrant waves, so that the resulting structure would be intelligible in terms of the forces that moulded it.

It is clear that both the system and the environmental pressures in Burlington were different from those in Chicago at the turn of the century and thereafter. The environmental pressure toward urbanization consisted of a gradually diversifying population coming in waves over a long period of time and settling into sub-communities, while relations to the older settlers were formalized without serious disorganization of the kind ordinarily associated with urbanization. This is a different pattern from the kind identified in Chicago, but still no less an instance of urbanization and susceptible to interpretation in these terms if suitable modifications in the theory are made. It is exactly this kind of application of the theory to divergent cases which establishes deeper insight into the processes involved. The Chicago pattern is simply one extreme instance of a generalized environmental pressure conceptualized in Park's theory of urbanization which requires extension to other and quite different communities, so that its different forms can be explored.

One might say the same for industrialization and bureaucratization. While Lynd shows the effects of industrialization on production and consumption patterns in a small Midwestern town, we know that these effects would vary significantly in a larger city even during the same time period. Absentee ownership which so polarized Newburyport had a much slighter effect on Muncie, as shown by the lack of any unusual hostilities toward the absentee-owned General Motors plant. Clearly, then, all of these environmental pressures can be better understood if examined in various community contexts and their effects related to their own variant forms as well as to the social structure of the community on which they are exerting an influence. *Under these circumstances, substantive propositions will always have a field-theoretical character in that they refer to the impact of a particular form of environmental pressure on the structure of a particular kind of community.*

It may prove helpful to block out some of the more formal aspects of the theory of community development growing out of the first three chapters. First, every community study must start with a description of the social structure as this has changed over a period of time. Whenever possible, attention should focus on aspects of the structure deemed problematic by the participants as well as by the observer. Earlier studies provide a set of models indicating possible structural patterns but they cannot be projected onto any unique situation without first testing to determine whether or not they fit. Park sensitizes us to the constellation of urban sub-communities found in Chicago in the twenties; yet this certainly cannot be taken as an accurate map of the sub-community constellation in any other American community, either now or then. Similarly, the Lynds' pictures of Muncie social structure, whether as two-class or three-class conceptions, must be applied with caution. They sensitize us to class sub-communities of considerable importance and indicate possible relationships between such sub-communities—especially differential power—which should not be neglected in any field work. Warner's stratificational model goes even further by suggesting that prestige, power, and economic position are not necessarily correlated and thereby calls attention to "old family" aristocracies as well as the "nouveau-riche." In addition, he points to the possibility of complicated interplay between ethnic status and the social hierarchy which demands exploration in each instance. All of these models imply special bases for structuring likely to be found in any American community. But the task of the field worker is to use these models to identify the actual structures in the particular community under investigation. In other words, specification of the social structure, though not undertaken in a theoretical vacuum, remains a creative task.

Once the picture of the social structure has begun to emerge, patterns of change should also be carefully noted. How did the observed structure come into being and what are the forces conducive toward its modification in determinate directions? Here the internal social system of each sub-community has to be investigated. Is it divided along generation lines into "orthodox" traditional practices upheld by the older members, in contrast to "heterodox" novel tendencies among the youngsters? What are the "problems" felt by the whole community as differentiated from those felt only by particular sub-communities and even by its segments?

This preliminary structural "sizing up" of the community and its sub-community constellation should pay close attention to directions of change. Ecological and demographic data can sometimes be helpful in this connection. Are there any coherent "neighborhoods" in the city and if there are do their residents so identify them? Can a pattern of distribution of disorganized behaviors be identified and if so how is this related to the sub-community constellation? It is important to remember that the social structures must always be determined by examination of institutional patterns and the problematic behavior fitted into this. Analyses in this theory always move through the several levels: the whole community, a sub-community, sub-community strata, and individual role-playing. Warner provides a different kind of problem-centered model in his analysis of the strike. Here, tensions in a specific social system, the factory, are interpreted as reflecting changes in the relations between various strata in Yankee City which, in turn, reflect the assumption of control by absentee owners as part of the nationwide bureaucratization of industry. It is very important to distinguish these various levels so as to identify the point at which any particular analysis falls.

Once the social structure has been provisionally charted, Park's theory of urbanization should prove an excellent starting point. It forces attention toward the elements making up the population and raises questions about their coalescence into sub-communities over the period of the growth of the city. Current disorganization can be studied in the context of the sub-communities in which it appears and the control mechanisms related to the behaviors they are supposed to contain. Novel sub-communities not anticipated by the theory should be carefully noted for further study. Again, the absence of any that might have been expected is equally significant and must be explained. Thus the decline of Bohemias in some American cities like Boston is as important as the proliferation of many different kinds of suburbs.

When the effects of urbanization have been assessed, our two other environmental pressures, industrialization and bureaucratization, should be inspected. Here it is difficult to know clearly whether one is studying the system or its environment, but this ambiguity haunts all "field-theoretical" interpretations. Though few communities today will still be undergoing the transition from craft to industrial economy that were discerned in Muncie, the Lynds give us the tools for analyzing industrialization. Except in isolated "backward" areas, industrial technique in production and consumption will already have taken a deep hold. Most jobs are mechanized and employ some degree of assembly-line specialization while most families depend mainly on mass-produced goods to provide the necessities and status insignia that make up life in our time. But this general view obscures the many pressures on community life arising from regularly occurring technological innovation in methods of production which destroy some jobs and create new ones. Perhaps the spread of automation heralds the next major phase of industrialization. Its effects on community life, both directly through changes in work roles and indirectly through the increased leisure it will make available, are incalculable. Consumption patterns are still being modified in line with the quest for an ever-expanding standard of living as the Lynds emphasized, and this, in turn, leaves the identification of consequent changes in every area of institutional life an open area for investigation. Perhaps the much-debated effects of television are a case in point, though recent studies seldom attain the sophistication of the Lynds' work because they do not locate the ramifying consequences of television in the context of *changing* total community patterns.

Middletown in Transition provides a valuable model for placing the community in the context of the historical period so as to pinpoint the effects of the decade experiences on the main patterns of change. The Lynds' contribution is not so much that they decided to study the effects of the depression, but that they studied its effects on the changes resulting from industrialization. Thus, the short-term (decade) historical context is introduced so as to maximize its analytic relevance. We learn that the depression did not diminish the impulse toward conspicuous acquisition but that it did create an awareness on the part of workers that large-scale agencies like trade unions and political parties could help in securing this. Similarly, any community study undertaken today would have to consider the implications of "prosperity," "recession," and the cold war as a context for observed internal changes.

Aside from calling our attention to additional bases for social structuring, Warner documents the effects of bureaucratization, the organizational concomitant of industrialization. He does this first by showing the effects of centralized absentee ownership on the attitudes of local workers who see the "outsiders" as alien and opposed to their interests. Then he shows how some degree of stabilization of the authority system in the factory is attained by acceptance of union bargaining. The traditional owner-managers in Yankee City have lost their power and wealth, and loss of their prestige is likely to follow. It is also likely that the stratum which Warner called "lower-upper" in 1935 has by now raised its status while the old "upper-uppers" continue downhill. Most American communities have probably accepted the fact of absentee ownership of large industry and the integration of corporate managers into community power structures is undoubtedly much more advanced. Furthermore, the legitimacy of such managers and owners could hardly be challenged on the same grounds as it was during the strike, if only because the circumstances in which factory manager-owners are also community leaders with the closeness to the workers that existed during the early period in Yankee City is no longer part of the American experience.

If we use these three sets of studies as points of departure, the foregoing should suggest an interpretive framework for examining other community studies as well as for formulating new ones. The conceptual model rests on the examination of change and assumes that urbanization, industrialization, and bureaucratization, as defined earlier, plot most of the key dimensions. Since the historical period of these three fundamental studies is confined to the twenties and thirties, we would expect that research completed today will deal with later stages. In field-theoretical terms, the problem today is as much one of identifying the contemporary stages of urbanization, industrialization, and bureaucratization through their manifestations as environmental pressures as it is of discovering the new patterns of community social structure that they call forth.

There is one underlying community trend to which all three of the studies refer. That is the trend toward increased interdependence and decreased local autonomy. Park referred to this process as urban or metropolitan dominance, and the Chicago sociologists did much to delineate its operations. Lynd noted the parallel phenomenon of dominance by centers of technological diffusion so that Muncie came to depend increasingly on Hollywood and New York for its entertainment, as it did on Detroit for industrial advance in the automotive plant. Warner dealt with this specifically in his discussion of the impact of absentee ownership, by which Yankee City shoe factories became the lowest echelon in a chain of command reaching upward to New York City. All of these studies during the twenties and thirties, then, show increasing standardization of community patterns throughout the country with agencies of nationwide diffusion and control acting as centers of innovation. Intimate life patterns become susceptible to standardized change as the mass media begin to inform each age group about new fashions and styles. Thus, the conditions for formation of a mass society were found in these previously examined social processes, even though more advanced forms had not yet appeared.

It is the study of problems like this for which the community theory being developed is singularly well adapted. For here is one junction point between macroscopic social theories and microscopic social research. The Lynds achieve such a juncture in *Middletown in Transition* when they show that the Muncie working class did not react to the depression as Marxist theories about class consciousness would have predicted. Their achievement, however, is not so much measured by this simple negative finding as it is in terms of their report on the institutional accommodations which preserved the dominant wish for mobility intact despite widespread unemployment. Since Marx's hypothesis referred to a national proletariat and not to that segment found in a small Midwestern American town, Lynd's findings can hardly negate any more than they could confirm this broad Marxist hypothesis. What they can do is give us some insight into the workings of the process whereby class consciousness is generated and shaped by focusing on its specific determinants in this particular context. If we could gather enough data on reactions to the depression in other community contexts, then we would have some conception as to the circumstances under which the Marxist form of consciousness appears as against those under which it does not.

This is the most clear-cut example and one peculiarly well adapted for showing the possible uses of community studies to clarify our understanding of basic social processes. The whole issue of social stratification in America, which is so central to social theory, can clearly be interpreted in terms of systems of community stratification as well as in terms of strata that presumably cut across the whole national social structure. Warner is as much in error when he projects the rather unique stratificational system of Newburyport onto the entire United States as are students of national stratification when they project some simplified "national" system onto the complicated communities and sub-communities of which our country is made up. The point here is not to discourage speculation at either level but rather to identify the kinds of problems that can best be handled by each as well as to see where they can fruitfully interpenetrate.

To carry somewhat further the argument that community studies can contribute to classical social theory, it is useful to look at the Yankee City Series in the context of Weber's theory of bureaucratization. Here we clearly have a transition in the factory from a traditional system of authority to a bureaucratic order. Warner's description of the problems involved in such a transition are of considerable importance in that they point up the difficulties in legitimating within the factory impersonal authority prerogatives that are not accompanied by equivalent status in the community. This has long been a problem of American business, and institutional advertising is one technique designed to cope with it in the society at large. Warner suggests that Yankee City managers will have to pay more attention to the status structure of the local community so as eventually to find places of some prominence in it. At the same time, they cannot become too responsive to local needs because this could interfere with the demands of profit-making when, as frequently happens, the interests of the two diverge. In any event, Warner has identified an important range of variation revolving about the orientation of absentee managers to local status structures. This would apply to non-economic as well as to industrial bureaucracies according to their need for local legitimacy.

Another study, Alvin Gouldner's *Patterns of Industrial Bureaucracy*, reveals the operation of many of the same factors. It deals with a gypsum plant that has been managed by local figures even though owned by a large chain. When top management decided to "rationalize" the plant instead of allowing it to function on a relatively informal basis, they brought in an "outside" manager who ran into serious difficulties with any innovations he introduced. Ultimately, the new manager's efforts at instituting rules to regularize work patterns led to a strike which Gouldner neatly analyzes in his companion volume *Wildcat Strike*. Even though Gouldner did not study the community in any detail, the two sets of studies provide interesting comparative data on the problems of bureaucratization in different contexts. Gouldner's discovery that even minor technological innovations (industrialization) were resisted vigorously when managerial motives were distrusted bears out Warner's findings. Furthermore, his finding that resistance to technological innovation is in direct proportion to its challenge to the status prerogatives accorded workers by the factory social system supplements Warner's finding that hostile responses were generated by working-class loss of community status. Both these studies enlarge our knowledge about bureaucratization considerably—Warner's providing tools for analyzing the interplay between the process in a specific social system and the local community in which that system is embedded.

Warner's description of the rise and acceptance of the trade union is quite important in that it follows Weber's suggestion that bureaucracies can be challenged only by equivalent centralized power structures. Contrary to Marx's expectations, this does not seem to enhance "class consciousness." The bulk of the workers are at the bottom in both hierarchies though there is reason to suppose that under most circumstances the union hierarchy will remain more responsive to their demands. They lose any sense of real relationship to their local community and remain only pragmatically identified with the union. Their conceptions of the workings of capitalism, even in response to the crisis of the depression, remain utterly parochial without the insight or even the sense of class solidarity that Marxists expected. Instead, they seem to have embarked on a collective effort to make their jobs secure and raise their standard of living through rational-bureaucratic procedures with the union as their vehicle.

There are, of course, many central sociological problems that fall outside the purview of the community study. An important instance would be the interpretation of national power like that offered by C. Wright Mills in *The Power Elite*. Similarly, the community study can hardly hope to throw much light on the origins of the environmental pressures which it takes for granted. While it can show the effects of urbanization, industrialization, bureaucratization, and even the depression in a specific context, it can hardly explain these supra-local processes themselves. Broad problems of social structure such as the dispute between "mass" and "class" theories can be dealt with only fragmentarily, though the evidence in the studies reported thus far suggests that both models apply in varying degrees.

Some comments on the nature of modern communities might help to clarify the scope of this theory of communities. Park distinguished between three aspects or levels of community— the biotic, the moral, and the spatial. "Biotic" refers to functional interdependence, "moral" to group identification and loyalty, while "spatial" implies distinctive location. In a simpler society, these three levels of community would probably coexist so that the group with which one lived would be the one with which one worked and to whom one's loyalties were paid. As Park well knew, these concepts could not be readily applied to community life in the city. Here, people often distinguish between their work community, their neighborhood, and their social circle. There is nothing that demands that these separate spheres of institutional participation include the same individuals or in any sense constitute a coherent social whole. Spatial neighborhoods may have no significant social meaning and true communal congeniality may exist between people scattered throughout a city.

Unlike role analyses which focus on a specific aspect of an individual's social participation, community sociologists always consider the full range of institutional participation as each aspect of it affects every other aspect. Thus, Lynd's analysis of the functional interconnections of the emerging industrial community in Muncie is distinguished exactly by this sense of institutional reciprocity so that changes in family consumption requirements can be seen as central for motivating industrial role playing. The individual's total life space—his modes of participating in major institutional areas—are always at the focus of attention. Changes in this life space receive careful study. The life history can provide essential data on the respondent's changing life space as he perceives it throughout his life. These changes have to be studied within the context of changes in his community systems. One central methodological problem of community studies lies in learning to coordinate these orders of data.

But it is impossible to take up all of the substantive and methodological problems raised by this theory in a single chapter. The word "toward" in the title must be underscored. It is included as a way of emphasizing the intention of the chapter—to indicate the elements of a theory and to block out an approach to its application. But it should also suggest the tasks that remain. There is a broad field open for anyone inclined to work out the systematic relations between urbanization, industrialization, and bureaucratization more carefully than has been attempted here. In the chapters that follow, the application of the theory to other community studies and to the interpretation of the problems of contemporary community life takes precedence over the development of systematic formulations. This choice rests upon personal predilection, as well as intellectual commitment, and should be construed as an invitation to sociologists with a stronger systematic bent to work *toward* developing a theory of communities in their own way.

PART II. DEVELOPMENT

INTRODUCTION TO PART II

THE five chapters in this part can be viewed as separate case studies—indeed, they deliberately highlight the distinctive findings and contributions in each instance. However, proper appreciation of the "individuality" of each case study should not interfere with recognizing its contribution to the theory. The interpretations always trace the disorganizing and reorganizing effects of urbanization, industrialization, and bureaucratization. Furthermore, viewed as a sequence, the five chapters plot certain structural trends in American community life from the twenties through the early fifties. They present the historical and conceptual material necessary for exploring the development, structure, and functioning of contemporary American communities.

Three chapters (5, 6, and 9) take up important urban subcommunities. Chapter 5 draws on William Foote Whyte's *Street Corner Society* to identify aspects of slum social structure that remain concealed if the slum is viewed simply as a disorganized area in the way that the Chicago sociologists tended to do. Chapter 6 turns to Caroline Ware's *Greenwich Village* for information about Bohemian social systems during the twenties. Finally, Chapter 9 uses three books on suburbs— *Crestwood Heights* by Seeley, Sim, and Loosley; *The Exurbanites* by Spectorsky; and the discussion of Park Forest in *The Organization Man* by William H. Whyte—to explore recent developments in this type of community. These three chapters are essentially studies of "natural areas" and enlarge our conception of the structure and functioning of the urban constellation.

Another important range of variation among communities which is strikingly exhibited, when a New England town like Newburyport is compared with a Midwest town like Muncie, is that of regional diversity in cultural and social structure. To deal with this more carefully, the studies of Southern communities, *Deep South* by Allison Davis, Burleigh Gardner, and Mary Gardner, and *Caste and Class in a Southern Town* by John Dollard are taken up in Chapter 7. Attention focuses on the effects of the caste system. Chapter 8 turns to another kind of caste system—that of the American Army in World War II. It tries to assess the meaning of the war as a "generation experience" by looking at the military communities in which so many American men spent the war years. Due to the lack of any conventional community studies, data are drawn from the first two volumes of *The American Soldier* edited by Samuel Stouffer et al. and Bill Mauldin's *Up Front*.

The discussion of suburbia assumes greater importance than the single chapter devoted to it might suggest. It is used in Part III as the basis for summarizing many important trends in American life. This is justified on the ground that an ever-increasing percentage of the population lives in suburbs of one kind or another, and a great many of those who still do not would do so if they could. One suspects that Park would have been as preoccupied with the suburb were he alive today as he was with the slum and the ethnic ghetto during his lifetime. It is a useful social laboratory in which to study disorganizing and reorganizing trends in mid-twentieth century America.

Due to the need for keeping the size of Part II within bounds, several excellent studies, including Warner's *Democracy in Jonesville*, August Hollingshead's *Elmtown's Youth*, and James West's *Plainville, U.S.A.*, had to be omitted. Everett Hughes's perceptive study, *French Canada in Transition*, throws light on all of the processes taken up in Part I, but his emphasis on the specific Canadian setting made it necessary to forgo the pleasure of including it as a case study in this book.

THE SLUM: STREET CORNER SOCIETY

NOVELISTS portraying city life have been as fascinated with slums as were the urban sociologists. In both instances, the preoccupation stemmed from an image of the slum as a place where the violent contrasts of city life found their sharpest expression. Slums were seen both as urban jungles in which lawlessness prevailed and, because of their association with tight little immigrant colonies, as the last stronghold of traditional intimate social life in the impersonal city. These two images—the slum as jungle and the slum as ghetto—still dominate both the sociological and the literary approaches to this sub-community.

Whatever the starting-point, slums are surely the most studied of urban sub-communities. Chapter 1, on Park, reviewed some of the research undertaken during the twenties, especially the work of Zorbaugh, Thrasher, and Shaw. These investigators provide a good preliminary conception of slum social structure, and its most disturbing feature—juvenile delinquency. Some of the relevant propositions in this connection have already been reviewed. The contribution of the Chicago sociologists lies in their recognition that delinquency is group-organized, or "gang" behavior, rather than the behavior of isolated individuals. Furthermore, they began to analyze the ways in which the social structure and values of the slum made the emergence of such gangs inevitable.

Unfortunately, most of their interpretations of slum life tended to emphasize its "disorganized" aspects. They were obviously aware that this was "organized disorganization," but their emphasis on the latter led to neglect of the former. While

the Chicago conception of urban disorganization forced attention to certain crucial problems of social order in large, rapidly
growing, heterogeneous population aggregates, it did tend to
define *some* of these problems too narrowly in terms of middle-
class values. Though ideological critics of this school have
pushed this valid point somewhat too far by claiming that
the entire concept must be repudiated because of its bias,
calmer reflection suggests that more can be accomplished by
subjecting it to critical inspection along several directions.

First, even the most vigorous critics would hardly deny that
many of the behaviors characterized by the urban sociologists
as disorganized were actually objective threats to social stability in the cities of the twenties. Crimes of violence would
be the most obvious instance. Other such analyses, however,
are more dubious—for example, taking a high or rising divorce
rate as evidence of disorganization. For, under certain circumstances, divorce supplemented by probable remarriage could
be seen as a new form of social organization peculiarly well
adapted to the conditions of urban life in which all too fallible
romantic impulses had become the basis for early marriage.

Secondly, it is possible that a given pattern might be considered disorganizing from the perspective of the whole city
while still an integral part of the social organization of a particular sub-community. In other words, the same item can be
functional for one unit while dysfunctional for another. Or, it
can be dysfunctional in some respects and functional in others,
depending upon the perspective adopted. Thus, the juvenile
"gangs" were clearly an essential component of the *organized*
social structure of the slum even though they occasionally engaged in activities that threatened their middle-class neighbors or, perhaps more correctly, they violated the values of
these same neighbors by virtue of their presumed evil influence on youngsters in the slum.

For a comprehensive picture of slum social organization,
less encumbered by middle-class assumptions, the study *Street*

Corner Society by William Foote Whyte is indispensable. It would be a mistake, however, to assume that Whyte's exploration of this long-established Italian slum vitiates either the findings or most of the interpretations advanced by Park's students about the slums of Chicago during the twenties. Instead, it is more profitably viewed as a case study of a well-established slum in which relations with the outer city and its control system have been stabilized. Eastern City—in this case Boston—was older and consequently its neighborhoods were undergoing much slower transitions than those in Chicago. At the time of Whyte's research, Cornerville had been solidly Italian for thirty years. It is not surprising that relations between the various sub-groups, as well as between these subgroups and the representatives of external authority, had been clearly defined. Yet one would hardly expect to find such clear structuring in New York's transitional Puerto Rican slums, where the police are sometimes compelled to patrol in teams of three.

While this context helps to explain the structural stability that Whyte found in Cornerville, the fact that his research was completed during the depression helps to account for the existence of "gangs" of unemployed young adults. Today there are few Cornerville gangs with their members in their mid-twenties or above who actually spend most of their time together.

Once having noted the particular historical context in which this study was conducted, it is possible to turn to the research itself for a portrait of an urban sub-community, the ethnic slum. Actually, it is far more relevant as a slum study than as an ethnic study simply because Whyte chose to concentrate his attention on the former. Even while noting their importance, he did not go into any detail about the functioning of the Catholic church or the Italian family. The advantage of his focus is that the distinctive slum institutions are highlighted and the structural interrelationships which transcend ethnic boundaries carefully

[121]

blocked out. Thus, the model presented here of patterned relations between corner-boy gangs, local racketeers, politicians, and the police would hold true to a considerable degree in slums with non-Italian ethnic populations. Furthermore, recent observations suggest that, with certain minor modifications, the picture presented by Whyte remains essentially accurate for Cornerville twenty years later.

Perhaps the most familiar and, in many ways, most important section of Whyte's book is that in which he describes two gangs —the Nortons and the Cornerville Community Club, led by "Doc" and "Chick" respectively. The rich picture of the structure and activities of these two gangs has become classic for the sociology of small groups and the sociology of leadership.

From the perspective of a theory of communities, the analysis of these "gangs" provides information about the relations between different kinds of slum gangs and about the processes of ethnic assimilation and social mobility. There is an underlying assumption to the effect that Doc's boys, the Nortons, and Chick's Community Club represent special types of social groupings formed to satisfy needs that emerge in any slum setting. The former represents the adaptation of corner boys, the latter of college boys, though both contain persons marginal to each group. It is clear that neither represents a "delinquent" gang in the conventional sense, even though the corner boys regularly engaged in technically illegal activities as part of their integration with the non-legitimate political and racketeering social systems which were really the only occupational outlets available. Through careful examination of these two gangs, each representing a different adaptation to the deprivations of slum life, many central features of slum social structure are illuminated.

In addition to the sustained comparison of the two gangs, Whyte's comparison of the gang leaders, Doc and Chick, is a brilliant analytic device. Here biography is turned into an instrument of sociological analysis so that Doc typifies the talented leader who remains within the social context of the slum sub-community, while Chick's path is that of the equally, or perhaps differently, talented individual who chooses to leave his home community to make his way in the larger society. The fact that neither apparently did continue on exactly the paths Whyte charted does not make his sketches any the less valuable—though it should force a reexamination to explain the later findings.

Doc's reluctance to strive for upward mobility and apparent willingness to remain loyal to his boys could have become political assets if he had been able to obtain sufficient financial backing to broaden his contacts beyond his own gang. His inability to do this was at least partly due to the depression, but one suspects that the very qualities which kept him so close to his boys were the ones that limited his range of influence. His was the gift for leading small groups and the evidence in this case suggests that this is not always the same as the qualities required for large group leadership. His uncanny insight into each of the members of the Nortons required close and prolonged contact.

Chick's club, on the other hand, had as its main purpose assisting the upward mobility of its members. Where the Nortons had no formal charter, the Community Club professed its dedication to the improvement of Cornerville. One met on street corners, the other in the settlement house. The fact that Chick's club had several members who attended college contrasted sharply with the casual employments of the Nortons. Parliamentary procedure would have been as alien to the latter as it was natural to the former. Chick's thrift is matched by Doc's free spending, and both must be interpreted in terms of their putative life plans and gang milieu.

The Nortons, led by Doc, and the Community Club, led by Chick, are always treated by Whyte in such a fashion as to illuminate the general features of gang structure and gang leadership patterns that they embody. It is true that neither of them is a really delinquent gang so that this book does not provide insight into the structures or leadership patterns of the kind of gang in which Shaw and Thrasher were primarily interested. Further research is required to establish the conditions under which slum sub-communities develop each of the three types: corner boy, college boy, and delinquent gangs.

With reference to leadership patterns, one might suspect that the "delinquent" gangs would be more authoritarian than the "non-delinquent" varieties, though Whyte certainly provides no information on this. It is doubtful that many other corner-boy gangs were as democratic or as skillfully led as the Nortons. Doc was clearly an exceptionally gifted and sensitive leader, and this was reflected in the structure of his gang. Indeed, it was Doc's awareness of the subtle aspects of leadership that provided Whyte with the best source of data on this subject. Chick, on the other hand, was much less atypical than Doc, and therefore his leadership style probably closely approximates that of other college-boy leaders. This study shows that important insight about sub-communities can be obtained from both atypical and typical individuals or groups. Whyte suggests that Doc's unusual talent did not mean that he led his group differently so much as it meant that he was conscious of performing as a leader operations that other corner boy leaders carried out without awareness.

The settlement house turns out to be one institution in the community toward which college-boy and corner-boy attitudes differ significantly. Chick, who frequently uses it as an auxiliary in his climb upward and outward from Cornerville, is well regarded by the social workers and reciprocates this feeling. Doc, on the other hand, distrusts the social workers and the settlement house. At one point, he accepts a job with them and achieves considerable success, but is eventually dropped because they ran out of funds. The social workers never quite accept Doc and have even more strenuous objections to the other corner boys. It is clear that settlement houses provide useful services for the boys headed out of the neighborhood and into the middle-class rather than for the stable slum dwellers.

In addition to their different feelings about the settlement house, it is clear that these two groups orient themselves differently toward education. The college boys are for the most part making considerable sacrifices to maintain themselves as students and therefore feel entitled to be respected. The corner boys have chosen to disregard schooling, many of them having dropped out at early stages. They resent the intellectual pretensions of the college boys as well as their efforts at assuming a middle-class style of life. The college boys are almost in the position of being "aliens" to their corner-boy peers although they must receive some cultural approval from the older generation, as well as from their fellow "aliens," the school teachers and social workers.

Whyte provides a neat contrast between the prospective political careers of Doc and Chick. The former refuses to sacrifice his ties to the Nortons and eventually withdraws from politics; the latter uses his leadership role in the Community Club to win the interest of certain non-Italian Eastern City politicians, finally going to work for the attorney-general in a minor staff position.

The important sociological feature of Whyte's use of Doc and Chick as types is his continual movement between their personal characteristics, their life plans, and their peer associations, on the one hand, and their relations with the institutions of Cornerville, on the other. He provides a model of sociological reporting in this connection. Thus, the Nortons are studied as a changing group whose place in the lives of the members declines as they assume new social positions, get jobs, and get married. Similarly, the Community Club is only a temporary instrument for its members, all of whom discard it when it no longer serves their purposes. Yet both clubs are shown to be essential agencies for the integration of their members with those aspects of Cornerville to which they are oriented. Though the processes and purposes differ in each instance, they are both agencies of socialization. Doc and Chick are then seen as social types arising out of and creating different gang structures which, in turn, lead to different relationships to the schools, settlement houses, rackets, and politics of the slum and therefore prepare their members for two distinct kinds of adult careers. Doc's career is presumably one within Cornerville while Chick's pivots around mobility into a non-Italian or higher class Italian sub-community.

This carefully wrought interpretive chain which links the two kinds of gangs to the social structure of the sub-community is as convincing as the system of institutional interrelationships that Lynd analyzed in the two classes in Muncie, and the probability of its recurrence as a basic pattern in urban ethnic slums is considerable.

Whyte also provides a detailed picture of the interrelationships of four common slum groupings which puts the above analysis into proper context. He shows how the rackets, the police, the corner-boy gangs, and the politicians formed a social system in which regular relations between the four were carefully maintained. During the depression there were few opportunities in Cornerville for any but the most fortunate to find respectable jobs. The college boys were in a tiny minority and young adults ensconced in permanent jobs of a legitimate kind were hardly more numerous. Men of Doc's age were forced onto the corner and once there could leave only by going into the rackets or politics. Later the army absorbed most of them, and today prosperity broadens the range of occupational choices considerably.

Whyte's picture of the functioning of the rackets in Cornerville is in sharp contrast to middle-class stereotypes. It is about the only business which is run with local people in reasonably high positions and as such offers a real opportunity for the more capable younger men in the neighborhood. The services offered, especially the "numbers" racket, appear to be an important way of cutting the monotony of slum life by providing a crack at "big" money for everyone. Racketeers are respected local figures depending for their success on good relations with the corner boys, who fill in as prospective customers as well as prospective employees. Generosity pays, so that Cornerville people find the friendliness of the racketeers quite satisfying when measured against the impersonal detachment of the alien business and professional people. Now that charismatic control of the rackets has given way to more regularized though not impersonalized arrangements, everything proceeds smoothly with the interests of all parties being carefully protected.

In the midst of all this, the Cornerville police, instead of being an impersonal agency of social control, turn up as willing collaborators of the racketeers, with the extent of their cooperation depending upon the amount of graft they can extract. So long as the racketeers keep within bounds, they are not ordinarily molested. When external pressure forces a clean-up, special officers are brought in from outside Cornerville to do the job. Politicians serve as intermediaries between the racketeers and the police. The interests of all these groups are satisfied when gambling operations are kept smooth. The system of reciprocal relations that Whyte describes so carefully is calculated to do this, and it even includes provisions for apparent enforcement when pressure is brought to bear by keeping a pool of "untouchable" officers available to serve this purpose. As Whyte observes, this system does not require "dishonest" police or "corrupt" politicians since the participants from both these groups can justify their contribution as the most effective method of keeping the neighborhood "quiet." They can do their respective duties and take in extra remuneration in the process. And since it is likely that persons operating within the system accept its values, there is little moral conflict: they are responding to the middle-class demand that the neighborhood be kept orderly while at the same time encouraging gambling, which satisfies essential needs of the residents.

From the point of view of urban sociological theory, the role of the police as described here is hardly that of a secondary control. They have clearly fitted themselves into the social structure of the sub-community in such a fashion as to achieve surface "orderliness" without disturbing one valued form of deviation—gambling. It is important to recognize that this is achieved by meshing with the local structure rather than by fighting to change it. Other more serious deviations are not, however, similarly tolerated, and it does not follow that because the police are informally tied up through the politicians with racketeers that they can easily or even at all pass over crimes of violence. Clearly, then, the workings of this particular agency of secondary control has to be seen in the specific context of the kinds of relations its agents establish with influential sub-groups in the sub-community in which it is supposed to operate.

Racketeering in Cornerville not only provides a good many jobs for local people; it also provides capital for legitimate enterprises. Corner-boy gangs depend on the racketeers for favors and reciprocate in kind. Playing leadership roles in these gangs is one steppingstone into the rackets as it is into politics. Whyte uses his technique of comparing two types of leaders in his interpretation of the roles played by Carlo and Tony in the Cornerville Social and Athletic Club. This analysis never probes as deeply as the comparison between Doc and Chick, but it does yield an interesting typology. Tony was the "outside" man in the club who maintained large-scale contacts with local politicians and racketeers. This meant that his capacity to maintain intimate ties within the club was limited, and that he had to cede inner leadership to Carlo. Both could limit the other's activities. Tony's outside maneuvers rested upon his base in the club, where Carlo's influence was greater. On the other hand, Carlo could accomplish very little without the help of Tony's outside contacts. So long as they recognized the legitimacy of each other's position and their need for each other, they could jointly hold the club together. Their political styles were different, with Tony depending more on impersonal accomplishment and Carlo on personal relationships. Tony's leadership depended on his difference from the corner boys, Carlo's on his similarity to them. Again, it is reasonable to suppose that other slums contain similar structures.

Clubs like the Cornerville Social and Athletic are the basic political unit of the neighborhood. Aspirant politicians have to obtain the support of these clubs and their leaders. With "repeating" a common practice, each club was good for many more votes than they had members. Young Cornerville politicians depended as much on the racketeers as on the political clubs with their corner-boy members. The latter supplied votes, but the former served as sources of funds without which campaigns could not be waged in a neighborhood like this. There are several patterns of interrelation between politicians, racketeers, and corner-boy club leaders, but the point to be emphasized is that every would-be candidate has to establish working relations with influential representatives of all three categories.

Like the rackets, politics seem to have become more bureaucratized during the depression: "Thus it appears that the New Deal helped to bring about a political reorganization whereby the localized organizations of ward bosses were to a great extent supplanted by a more centralized political organization headed by the United States senator, with the congressman next in line, and the ward politicians assuming more subordinate positions."[1] Cornerville is one of the least

[1] Whyte, William Foote, *Street Corner Society*, The University of Chicago Press, Chicago, Illinois, 1943, p. 198.

politically apathetic neighborhoods ever studied. Politics are part of daily life and depend on the same value systems. Cornerville political leaders command personal loyalties according to the same kind of reciprocal arrangements, here resting upon the exchange of services for loyalty (and votes), that tie together all of the groups in the slum. Politicians perform innumerable special services, meeting needs that could not otherwise be satisfied, in return for votes and support. Jobs are obtained, court charges "fixed," loans advanced, and even civic improvements occasionally accelerated through this network of reciprocal obligations. Alongside of this, organized political expression in middle-class neighborhoods, such as community improvement associations and parent-teacher organizations, appear remote from the main life concerns of residents.

From the perspective of the theory of communities, Whyte has presented an analytic picture of slum social structure and functioning. He shows it to be tightly organized on the basis of mutual interdependence between corner gangs, the political organizations, the rackets, and the police. Since he paid little attention to the distinctive Italian institutions—the church and the family—the role of ethnic variations is not carefully worked out. He suggests that the acceptance of gambling has Italian sub-cultural sanctions as well as roots in the deprivations of slum life but he does not follow up this lead or any of the others thrown out. It is worth noting again that some of the significance of the study actually derives from this limiting of his focus.

We can easily identify the contributions that Street Corner Society makes to Park's theory. As was noted, both the Chicago conceptions of disorganization and that of the relations between primary and secondary controls in the slum must be modified. Cornerville does not suffer from a simple deficiency in "organization" or from a lack of primary controls.

Whyte's contribution is to show the specific kinds of organization that arise in the stable slum based on primary relations and encompassing many of the local representatives of the secondary control system. The police are both compelled and motivated to tie in with the local structure, while those agencies which refuse to do this actually limit their effectiveness. Thus, the school and the settlement house exert only peripheral influence on the community so that their contribution toward assimilation is confined to persons already searching for avenues of mobility. Whyte never discusses the effects of the mass media as standardizing influences, but this may have been due to the fact that mass entertainment was far less important than gang activities because of the vast amount of leisure and scarcity of funds prevalent during the depression.

Substantial modifications in Park's theory are suggested by the apparent lack of criminally delinquent gangs in Cornerville. This really is the basis on which Whyte defines the neighborhood as organized (in contradistinction to disorganized)—though around different values from the middle-class world. Many of the characteristics of Cornerville, especially the stability of social structure and special relations to control agencies, can be explained if this particular slum is seen as a phase in urbanization. A sort of cycle of slum growth can be identified so that the "disorganized" slums of Chicago in the twenties are an early phase and this "stable" slum in Boston in the thirties a later one. Perhaps the distinction should be between "nascent" slums and "established" ones, with Cornerville constituting an extreme case of the latter. Clearly such factors as recency of settlement, homogeneity, likelihood of invasion by other ethnic groups, and similar aspects of the total urbanization picture will affect the stability of any particular ethnic slum. Whyte is aware of this total context but never explicitly fits his case study into it.

Industrialization in Cornerville takes devious forms. Thus,

racketeering has been broken down into specialized tasks. According to one informant: "The racket is organized just like a big business. Everybody has his own job to do. They got the con men, smart talkers—they can convince you that black is white. Then there's the muscle men. They muscle in to take over a business. There ain't much work for them now. There's the strong-arm men, they protect the business when it gets going. There's the killers. And then there's the bookkeepers, because they got plenty of accounts to figure every day. All them men get paid every week, and maybe for some of them there won't be no work for fifty-one weeks out of the year, but for that other week, there's plenty they got to do. You see, it's not like you see in the movies. Only a few of them are killers."[2] Certainly this is a highly specialized operation and the job breakdown is undoubtedly even more complicated than this suggests. Oddly enough, the movies are beginning to catch up with this social change, although the old-style gangster film still has its devotees.

More importantly though, the cyclic relation between industrial production and industrial consumption has affected Cornerville. Whyte's final interpretation of the social structure pivots around the problem of blocked mobility. Slum dwellers may take over the consumption preferences promulgated by the mass media, but they have no way of getting access to the advertised products except through "making good" in the rackets or politics. Some few can move up out of the slum as the college boys hope to, but this is a hazardous path often demanding repudiation of ethnic ties. The entire extra-legal structure of the slum is seen as an adaptive response to deprivation of access to the legitimate means for achieving mobility in the society at large.

The same informant who described specialization in the rackets to Whyte went on to illustrate aspects of bureaucratization: "Suppose you start to write numbers and back them

2 *Ibid.*, p. 120.

up yourself. When the office finds out about it, they will send
a con man around to see you. He's a man that can make you
believe that black is white, the way he talks. He says it would
be a good thing for you to join the protective association and
become a 50 per cent man if your collections are large enough.
They would protect you in case of a shakedown. You would
handle your own police protection, but if you can't reach
the cop on your beat, the office will handle that for you. He
tells you about the advantages of belonging to the organization and impresses on your mind that you better join. If you
don't listen to reason, you get a warning, and if you still don't
come in, you get beaten up. But nobody is going to hold out
on them anyway."[3] Large-scale organization triumphs here as
it did in Newburyport. Even the transition from charismatic
to managerial leadership is reenacted. Whyte compares the
old racket leader, Mario, who used his fists and his guns to
blast his way to the top, with his successor, T.S., an innoc-
uous person who demands that his organization keep violence
to a minimum. Mario and Rufus Choate might not have gotten
along well together, but they would probably have understood each other better than either would have understood
the absentee managers of Newburyport.

Viewed as kinds of bureaucracies, the rackets and political
machines of Cornerville have some interesting differences
from their more legitimate counterparts. Both depend heavily
on the good will of their clients and even the top leaders of
both maintain some contact with the leading representatives
of this clientele. Even though these organizations have some
of the standardizing, rationalizing, centralizing trends found
among businesses in a market economy, their situation forces
them to respect the traditional values of this sub-community
in a fashion not required of the absentee owners in Yankee
City. Trust and gratitude remain central to the racketeering
and political operations so that the organizations in both

[3] *Ibid.*, p. 121.

cases strive to mesh with the informal groupings of the neighborhood as closely as possible. They cannot survive without doing this. There seems to be more genuine correspondence of values and goals between these two Cornerville bureaucracies and the local residents than exists between the absentee-controlled shoe factories in Yankee City and the ordinary citizens and workers of that New England town.

This discussion of *Street Corner Society* shows the contributions that a study of a single strategic sub-community can make to the theory of communities. By presenting a detailed picture of the slum social system which binds racketeers, politicians, the police, and corner-boy gangs in a network of reciprocal services and obligations, it modifies the whole conception of "slum disorganization." Assumptions about the role of secondary control agencies like the schools, settlement houses, and the police are revised. Whyte's comparison of the life styles of Doc and Chick provides insight into alternative patterns of mobility and success within slums. And, finally, our conception of the workings of urbanization, industrialization, and bureaucratization has been expanded through exploration of the details of this case.

BOHEMIA: CREATION, NEGATION, FLIGHT, AND REVOLT

THOUGH hardly less fascinating than its next-door neighbor, the slum, our Bohemias have never received as much attention from American sociologists. This neglect is unfortunate, since in this special sub-community certain basic community processes can be seen in sharpest relief. There is only one piece of full-scale field research, Caroline Ware's *Greenwich Village, 1920-1930*, to counterpose to the myriad studies of slums in Chicago and elsewhere. Luckily, sociologically-oriented literary historians and critics have taken up some of these problems, with the result that a book like *Exile's Return* by Malcolm Cowley is helpful for interpreting the social and cultural structure of Greenwich Village during this period.

Bohemias present some unusual problems as well as opportunities for the student of communities. Because they are not as ubiquitous as slums, only the very largest cities have clearly defined neighborhoods that can be so characterized with any certainty. For that matter, in any cities besides New York, Chicago, and San Francisco, the would-be student of Bohemia has as his first and by no means easiest task that of finding the sub-community. In smaller cities it may well be residentially dispersed, though certain key institutions like universities, art stores, jazz night clubs, and foreign-film theaters can serve as preliminary clues.

Since New York's Greenwich Village has long been the prototype of American Bohemias, a study of its particular characteristics should yield some clues to the others. Even so, it would not be surprising to discover that cities differ in the kinds of Bohemias to which they give birth. Boston's Beacon Hill area, part of which has become Bohemianized, is far more genteel than its New York counterpart, but then it is also less commercialized. Clearly the place that any Bohemian sub-community holds in a particular city must be studied in the context of that city's development.

Caroline Ware's book contains a great deal more of value to the theory of communities than its title might suggest. Greenwich Village is a complicated urban neighborhood containing within its boundaries sizable Italian, Irish, and Jewish sub-communities as well as the Bohemians who have brought it fame. There is almost no social contact between the "Villagers" and the ethnic "locals"; the two groups live side by side under conditions of total anonymity. Indeed, one of Ware's most carefully documented observations is that this section of the city is not a social neighborhood in any sense of the term, but instead exists as a series of separate "social worlds," each isolated from the other except under the most unusual circumstances.

One of the distinctive contributions of this study is its picture of the mechanisms of segregation through which such utterly different sub-communities can somehow coexist side by side. Naturally there are certain points of regular contact, like public schools, settlement houses, and public recreational facilities. Efforts at organizing the whole neighborhood to take actions on issues presumably affecting the interest of the various sub-groups were usually no more successful than the efforts that Zorbaugh reported with the near North Side of Chicago which was equally polyglot. Each of the ethnic groups applied negative stereotypes to the others, but all joined in condemning the "Villagers" even more vigorously than they did each other: "No division within the ranks of the local people or in those of the Villagers compared in completeness with the gap between the two. Though an Italian might not speak to an Irishman or a business man might not associate with a laborer, each was conscious of the other and lived his life in terms of the presence of the other. The laborer looked with envy on the prosperous man and would emulate him if he could. The business man noticed the labor sufficiently to feel his superiority. At least within very wide boundaries, certain common standards of values and commonly accepted patterns of behavior obtained."[1] Villagers aside, there was a common "code of the streets" which seems to be an urban working-class code adjusting differences between ethnic groups in close propinquity by channeling their hostilities along gang lines while introducing the younger members of each group to toughness, gambling, and a sex code that protected the females by confining their "deviant" acts to outsiders.

Because the major contribution this book makes to our theory is its interpretation of Bohemia, Ware's valuable discussion of the social systems of these ethnic groups in this composite urban neighborhood must be passed over quickly. It will have to suffice if we remark that its adaptive social system is different from that in Yankee City, where the several ethnic groups have places in an over-all community-wide status system, and equally different from the ghetto experience in which each group leads an hermetic existence. Ware hints at some of these differences in her comments on the Jewish minority which lives scattered among the Italians: "At an earlier age than the Italian or the Irish child, each of whom started his career in an environment predominantly composed of members of his own group, and much earlier than the Jew of a Jewish neighborhood, the Jewish child of Greenwich Village learned that to be a Jew was to be different. Instead of adhering closely to the pattern of behavior which he found about him, he was taught from the first that he must act differently, and as soon as he began to notice the conduct of his

[1] Ware, Caroline, *Greenwich Village, 1920-1930*, Houghton Mifflin Company, New York, 1935, p. 125.

II. DEVELOPMENT

neighbors he realized that their rhythm of life differed from
his—the Sabbath food, Christmas which was nothing to him
and the High Holidays which were nothing to them. Whatever his later experience, the Jew raised in this community
had had impressed on him in his early years a consciousness
of difference which the Jew of the Bronx did not have."[2]
No other study examined thus far shows more clearly how
ethnic ghettos affect each other while maintaining social distance in urban contexts. Ware heightens our sense of the
different meanings of being a Jew, for example, according to
the position and character of the sub-community in which
one lives.

Even after we lump these diverse ethnic sub-communities
together as "locals," they still present sharp contrasts to the
Villagers. Their mutual comprehension and orientation, even
when hostile, contrasts with their hostility toward the profoundly "alien" Villager world. Where all of the ethnic groups
accept the American dream of success as it is defined in monetary terms, its rejection is a key assumption in the Bohemian
ethic. Another point of conflict was the Villagers' rejection of
"bourgeois morality" coupled with their experimental attitudes
toward sex and marriage. The locals on their side were tied
up with such characteristic slum institutions as gambling, racketeering, and political machines, none of which was acceptable
to most of the Villagers. Indeed, it is a tribute to the effectiveness of urban segregation mechanisms that there was so small
an amount of overt conflict between two groups differing
sharply on many vital matters.

A deeper understanding of the relations between Villagers
and locals rests upon some conception of the development of
the whole neighborhood. Caroline Ware describes this in considerable detail in her chapter "Old Houses and New People."
Greenwich Village had been called the "American Ward" by
New Yorkers up until the turn of the century, mainly because

[2] *Ibid.*, p. 139.

it had successfully resisted immigrant invasions and retained a sort of small-town atmosphere. The growing city had by-passed it largely because the irregular diagonal blocks interfered with through avenues. But the forces of urbanization could not be held off completely, and a large Italian colony, the backwash of "Little Italy," moved in, bringing tenements and slum social patterns. The Irish already had roots but had not changed the dominant quaint Old American tone of the neighborhood. This invasion was resisted and housing reclamation projects to prevent spreading deterioration managed to preserve islands which persist to this day. The subway further opened the neighborhood to the rest of the city, but stabilization was achieved in the early twenties when ethnic invasions ceased because the Village became too expensive for immigrants and not stylish enough for those seeking suitable locations for second settlement colonies.

During the early period of its existence as a small-town island within the city, a tourist trade had been established on the basis of quaintness and historical associations. The ethnic influx threatened this at first, but the Bohemian influx re-established it in a different way. The early Bohemians resembled the Old American inhabitants in that they too were reacting against modern urban-industrial America as represented by the stifling commercial atmosphere of both the small towns and the big cities. These early discoverers of the Village usually cherished the traditional American craft-individualist values that were then under attack from all sides and hoped to create or preserve a social system in which these values could be upheld. Reacting against bourgeois "civilization" mainly in terms of pre-industrial standards, they stood for serious pursuit of artistic or creative craft-like endeavors uncorrupted by financial constraints. These men made Greenwich Village into a community, both real and symbolic, offering a promise of release from narrow middle-class values and the opportunity to pursue other life paths freely.

[139]

Ware identifies a second group of Bohemians, who lacked the ties to an older American tradition that sustained the Old Bohemians but were bent on establishing their own tradition based on the creative pursuit of art and sex for purposes of self-realization and self-expression. These were the New Bohemians, striving to revolutionize both their private and public lives along lines apposite to the dominant commercial ethic. They were almost as critical of the Old Bohemian traditional Americanism as they were of the middle class. Their search for individuality carried them into ever new forms of avant-gardism. Many in this group had serious aesthetic and personal standards, but others had emerged out of their headlong flight away from middle-class America with little more than a negative identity. This last group, whom we shall call Anomic Bohemians, knew what they were against but had only vague ideas as to what they supported. Instead of engaging in creative artistic pursuits, they tended to pick up the "newest" in each line of endeavor so that "shocking novelty" became their main value. They drifted along without ever establishing institutional arrangements like the craft styles of the Old Bohemians or the school and family patterns of the New Bohemians. Still, they were in genuine revolt; their tragedy stemmed from their difficulty in attaching themselves to any lasting program through which this revolt could be expressed.

Together with these three groups, a fourth began appearing in the Village during the twenties. As rents in the area went up because of a housing shortage, many of the poorer artists were forced to move to cheaper neighborhoods, and their place was taken by younger respectable business and professional men who liked the informal atmosphere of the Village. These were really Part-Time Bohemians whose work lives were utterly conventional and whose dedication to revolt affected only their leisure. Not without some justice, Ware calls these people Pseudo-Bohemians. More clear-cut cases of Pseudo-

Bohemianism were those individuals who exploited Bohemian poses to market products and services to gullible tourists.

Conceptualizing Ware's distinctions with some modifications, we may conclude that during the twenties there were three types of genuine Bohemians—the Old, the New, and the Anomic, with two types of Pseudo-Bohemians, the Part-Time and the Commercial. These distinctions derive their importance from the insight they throw into the processes of change in the Village. During the twenties it was already clear that the Old Bohemians were disappearing while the Part-Timers were entering the Village in ever-greater numbers. The New and Anomic Bohemians, unable to afford the exorbitant rents, were also giving way. As the tourist appeal of the Village mounted, it drew in more Commercial types, whose only interest was in exploiting the more sensational aspects of Bohemian life. Even during the twenties the trend was toward Pseudo-Bohemianism at the expense of the genuine varieties. Rather than serving exclusively as a haven for artistic and sexual non-conformists, the Village was becoming a bedroom community for junior executives, young professionals, and office workers who remained there only until they had families, after which they quickly shifted to more conventional surroundings.

Ware is careful to note that some of the social institutions for which the Village was most renowned were developed by the Pseudo-Bohemians, who depended upon hangouts and studio parties to meet their social needs. Their preoccupations —sex and drink—were the preoccupations associated in the public mind with Village Bohemians. Their hangouts and wild parties drew people from all over the city in search of "adventure and escape":

"The first group of Villagers had been made up of individuals of exceptional independence, who had faced social problems with earnestness and had sought positive solutions. When the community had come to contain a large proportion

[141]

of persons of ordinary caliber whose position reflected the social situations from which they had come more than the personal quality of the individuals, the negative desire to escape took the place of any positive quest, and social earnestness gave way to a drifting attitude. At the same time, the actual expression of repudiation became more extreme as the conduct which had constituted social defiance at one time became commonplace a few years later. The disappearance of smoking as an issue, the spread of drinking, and the passing on from free love to homosexuality were only the more obvious of the manifestations which were successively adopted to mark the outposts of revolt."[3]

These Pseudo-Bohemians concerned with the external trappings of defiance found themselves driven to ever-greater extremes as their old patterns lost any shock value. Clearly, a kind of inner dynamic is here operating to drive a defiant Bohemianism, whether pseudo or genuine, to its outer limits. Movements such as Dadaism are an instance on a genuine, though possibly anomic, level of this pressure toward extremism. For better or worse, it appears that the movements of the twenties all but exhausted the range of extreme possibilities, leaving the would-be innovator in the realm of deviant Bohemianism with a real problem.

Discussing the departure of the talented artists, whether New or Old Bohemians, Ware notes their dissatisfaction with the quality of the invading groups. In addition, however, their own life circumstances changed so that the Village became less attractive in comparison with the Left Bank or even the suburbs. Rising rents were also involved: "As the Village became better known and more generally sought as a place of residence, it lost the cheap rents which made it particularly attractive to artists and writers. At the same time, many of those who had struggled in the Village in poverty and fraternity had become famous and sufficiently prosperous to live

[3] *Ibid.*, p. 238.

elsewhere with greater comfort and independence. Thus, those with the money to stay could also afford to leave and those without money could not afford to remain."[4] One wonders if this does not describe the pattern of any Bohemia—a pattern by which the deterioration of genuine Bohemianism almost naturally follows a successful creative period.

Still, almost all the Villagers showed a configuration of values involving dedication to art, sexual liberality, and disdain for the pursuit of wealth. This configuration stood against the temper of the twenties as represented in Lynd's *Middletown*. Pseudo-Bohemians may have paid it only lip service, but even that required some kind of break with their home communities. Furthermore, this configuration of values with its underlying emphasis on creative self-expression provided the context for a set of institutional patterns that dominated Village life just as the rackets-politics-gang constellation dominated the slums.

These institutional patterns were the "progressive" school, the "equalitarian" marriage, and the "creative" career. Ware never clearly indicates the extent to which each of these patterns stemmed from the Old Bohemian repudiation of commercialized America, but it does seem likely that the New Bohemians played a greater role in their institutionalization, since they fall closer to the values of this last group. Needless to say, the Anomic Bohemians were not concerned with these aspects of life while the Part-Timers fitted in or not at their own discretion. Clearly this three-sided pattern was most important for the married Villagers of all categories.

The conception of "creative" careers was closely linked to the "equalitarian" family, since it rested upon the assumption that both husband and wife would have them. Household slavery was to be abolished or, if that could not be managed, then distributed equally among adult members of the family. Some of the contradictions in this kind of family arrangement

[4] *Ibid.*, p. 241.

are summed up by Ware in the following passage. It is quoted in full because much of it still holds for the contemporary suburban family, though the differences in sub-community context naturally remain significant:

"The families who sent their children to the local progressive schools, and more who had no children, attempted to carry into their married lives the principles of individual development and a disregard for such traditional standards as they intellectually condemned. Here again, inconsistency placed them under a strain which fell primarily on the women. In these families, the idea that the man should furnish the full support of the family had been discarded and the wife contributed as a matter of course to the support of the home. Subordination of the interests or personality of either to the other or to the interest of the family as a whole had also been scrapped in favor of the assumption that each was to retain the full measure of his or her individuality. If a man's creative work was incompatible with his contributing even a pro-rata share to the joint household, the household rather than the work should suffer. In these families, however, the idea of the home as a unit and as a center was not abandoned. In fact, the standard for the home which these couples struggled to achieve was usually set by the comfortable and attractive surroundings among which they had grown up.

"The effort to maintain such homes under city conditions placed a heavy burden on the family budget and especially on the capacity of the women. Often forced to do a large part of the housework because they could not afford maid service, while at the same time they went out to work and carried the supervision and care of the children besides, these women did not find it easy to impart to the home the serenity and sense of security which they felt it should possess. In embracing the newer educational and psychological doctrines, they expanded their concept of the responsibilities which they owed to their children, but were thereby still further pressed

[144]

in time and money to provide the opportunities, guidance, and care which they thought essential. They usually felt unable to take advantage of public facilities and had to pay for recreation which, in a country community, would have been provided by nature. Added to all this was the necessity of making a harmonious group out of intensely individualistic and self-expressive instead of disciplined and subordinated individuals. Saddled thus with heavy economic burdens, with constant calls for activity, with the personality problems of themselves, their husbands, and their children, they constantly faced the final insecurity of their whole marriage state. Because of the assumption that marriage was an arrangement terminable at will rather than a permanent union, they could rely on neither moral nor legal backing to help them hold their husbands, nor—and here was the crucial point of uncertainty in the whole situation—could they be quite sure that their husband really preferred their type of independent companion to the women who made a business of love and home and for whom they might any day be deserted. It was not surprising that one of the experimental schools found evidence of more or less strain in the homes of half of its children."[5]

Here, in the problems of the "experimental" family in the twenties, are embryonic versions of many of the most pressing problems of the suburban American family of the fifties.

The burden of difficulty clearly falls on the woman. Her role is fraught with contradictory demands yet devoid of compensatory rewards. She ends up doing several different jobs at terrible cost to her nervous system. Apparently, premarital "free love" entailed similar sacrifices, since even the enlightened male Bohemians of the Village were likely to turn against their own equalitarian standards when it came to choosing wives.

From the above quoted passage, it is clear that new responsibilities were added to parenthood, especially responsibilities

[5] *Ibid.*, pp. 259-260.

for the psychic state of the child. This burden has partly shifted to the "progressive" school but doing so added its own new strain to the already overtaxed family budget. Furthermore, these schools, with their emphasis on self-expression and personality development, reenforced the super-individualism which threatened family life. Logically, at least, it would seem that in this kind of institutional constellation, where both parents are committed to creative self-expression in the home and in their careers, the values inculcated by progressive schools would have maximum effect. In reality, all the inner conflicts besetting the experimental family, including parental concern over the effects their own insecurities were having on their children, had a way of inhibiting genuine creativity as much as the more authoritarian families in which the parents had been raised

One of the problems faced by Greenwich Village Bohemians was that of staying ahead of the rapidly changing society in which they lived. Ware tells us that even during the twenties there were wild parties in Brooklyn that would have made the most hardened Villager blush. The rate of change has accelerated since then with the result that life styles once deemed experimental and avant-garde have been incorporated into the everyday round by conventional conservative sub-communities. The ideals of the experimental family and progressive education have been absorbed by the middle class, leaving contemporary Bohemians hard put to distinguish themselves from the society at large. Actually, between the time of Ware's study and the present, Bohemian innovations in the Village were absorbed almost as fast as they appeared. An example of this is the readiness that advertisers display to seize upon avant-garde artistic innovations for their own commercial purposes. If Kinsey and modern advertising can be accepted as evidence, free love is now a sub-rosa pattern for many adolescents, while heavy drinking has become more properly a symbol of belongingness than social defiance in

[146]

many sub-groups. Both Lynd and Park have noted a general widening of horizons due to the mass media, so that this becomes a specially intriguing instance of a more widespread phenomenon. It is interesting to observe that the aspects of Village life presented to the wider public were its more sensational rather than its more serious phases. Pseudo-Bohemians received more attention than the genuine Villagers.

One of the few points of contact between locals and Villagers were the speakeasies. These were usually run by Italian gangsters, even though alcoholic beverages of various sorts were made and sold by almost every Italian shopkeeper, no matter what his ostensible business. The speakeasy operators had little respect for their Villager customers simply because the latter were too free with their money. Italians respected hard bargainers and were appalled at Villager willingness to pay double and triple the amounts they paid for tenement apartments hardly different from their own. Contrasting attitudes toward sex and family life brought their own set of conflicts whenever representatives of the two groups were forced into contact of any duration. However, Ware did notice that these contrasts were becoming less sharp as the ethnics became more Americanized and some of them adopted middle-class styles of life.

Since it is supposed to be a protest against middle-class values, Bohemia is one sub-community in which the social structure has to be studied in relation to the dominant values of a given period. Ware shows that although there was a core of values both negative and positive to which Villagers in the twenties adhered, the depth of their dedication to these values differed. The growing group of Part-Time Bohemians suggested that some kind of reconciliation between Bohemians and the bourgeoisie was in the offing. Malcolm Cowley succinctly summarizes the terms of this reconciliation in the following passage from his book, aptly titled *Exile's Return*:

"It happened that many of the Greenwich Village ideas

proved useful in the altered situation. Thus, self-expression and paganism encouraged a demand for all sorts of products—modern furniture, beach pajamas, cosmetics, colored bathrooms with toilet paper to match. Living for the moment meant buying an automobile, radio or house, using it now and paying for it tomorrow. Female equality was capable of doubling the consumption of products—cigarettes, for example—that had formerly been used by men alone. Even changing place would help to stimulate business in the country from which the artist was being expatriated. The exiles of art were also trade missionaries: involuntarily they increased the foreign demand for fountain pens, silk stockings, grapefruit and portable typewriters. They drew after them an invading army of tourists, thus swelling the profits of steamship lines and travel agencies. Everything fitted into the business picture."[6]

. Just as the cluster of "experimental" family, "progressive" school, and "creative" career could be absorbed by the burgeoning suburb, so the very values on which these had been founded were harnessed to facilitate the growth of the exact social forms against which they were supposed to be a desperate revolt.

But this was not the only fruit of the Village in the twenties. As Cowley shows, the established tradition of literary and artistic innovation was maintained at least during the early years, with two-way traffic between New York and Paris. The larger problem as to why this should have been a period of such consequential creative development cannot be dealt with here. Certainly though, the existence in Greenwich Village of a social system in which artistic and literary endeavors were supported and taken seriously must have played some part. That this system had already begun to deteriorate in the early twenties is also beyond question. As Gilbert Seldes

[6] Cowley, Malcolm, *Exile's Return*, The Viking Press, New York, 1951, pp. 62-63.

demonstrated in *The Seven Lively Arts*, even the popular arts —movies, comic strips, and the rest—were affected by the innovating spirit. It is likely that the influence of the Village played some role in this upsurge or perhaps both were part of a great cultural loosening caused by the solvents of urbanization and industrialization during these early phases.

From the standpoint of a general theory of communities, Bohemias perform some exceedingly important functions. Their disappearance, as heralded by Zorbaugh, is a matter of serious concern: "As the city expands, the slum moves on. Towertown, The Village, bohemia, however passes out of existence because it no longer has a role to play in the life of the city. As the outlying areas of the city become increasingly mobile, as contacts become increasingly secondary and anonymous in the great areas of apartment house life, group sanctions disintegrate, unconventional behavior is tolerated or ignored, and the 'radical' no longer faces the necessity of seeking refuge from community opinion. The bohemian way of life becomes increasingly characteristic of the city at large."[7] Zorbaugh was, of course, speaking about Chicago's Bohemia, but his impression coincides with the general drift of Ware's book. Zorbaugh never drew explicit distinctions between genuine and Pseudo-Bohemians but he notes the same dichotomy in Chicago even though he never interprets either type as subtly as does Ware. He too notes the diffusion of Bohemian patterns throughout the whole city, but his conclusion that this eliminates the necessity for any such sub-community does not necessarily follow.

During the twenties, it seems clear that the "villages" in our cities served to shelter deviants of all kinds who could find no place in the world outside. Some of these deviants —the creative artists and writers—made important contributions to our cultural life. The institutional innovations nour-

---

[7] Zorbaugh, Harvey W., *The Gold Coast and the Slum*, The University of Chicago Press, Chicago, 1929, pp. 103-104.

ished there in education and family life and the demand that work be creative were important correctives to the main drift in the twenties and important social building blocks for what was to become the main drift of the fifties. If Bohemias have actually disappeared, this is a matter of considerable consequence if only because community laboratories in which innovations are conceived and tested seem even more necessary today than thirty years ago.

The trend toward greater tolerance anticipated by Zorbaugh has suffered serious setbacks so that would-be deviants still need protection. It can be argued that the threats to creative artistic and literary careers today are greater and even more subtle than during the twenties. During the thirties, Greenwich Village drew an increasing number of political radicals who combined socialism with an interest in modern art and sexual liberalism, though the órthodox Communists in this group were forced to repudiate the last two and ignore the first when the Soviet dictatorship established its official line. The forties filled the Village with curiosity-seeking veterans, some of whom apparently made homes there in hopes that they could recapture the post-World War I ferment. Unfortunately this could not be revived artificially. Life had become far more complex, the stabilizing influence and leadership provided by the Old Bohemians was missing, and the fleshpots of the communications industry too easily accessible.

Cursory inspection of the Village today suggests that it is still a haven for adolescents in search of "experience." Yet serious artists still choose to live there, as do many university professors. Side by side with the inevitable grey-flanneled junior executives, and probably wearing the same uniform during business hours, the artists and intellectuals of the Village enjoy a slower and more informal life than is possible in most other parts of Manhattan. But distinctively Bohemian or even non-conformist elements are not prominent. Perhaps the homosexuals stand as the last clear-cut deviants, but it

[150]

appears that the conformity demanded within this tight little world is as exacting, if not more so, than that which prevails in more conventional sub-communities.

In an era of comparative prosperity such as the fifties have been thus far, it is hard to oppose the dominant values and ruling groups. A prosperous and amorphous America does not offer the same striking target that the narrowness and conventionality of small-town life around the turn of the century provided. It is difficult to continue advancing with the avant-garde when there does not seem to be any really new places to move toward and no one forbidding you to go there even if there were. Protest takes highly individualized forms so that the lone suffering of existentialist philosophers or novelists provides little basis for the creation of new social patterns, in part because its starting premise is that no such basis exists. An equally individualized solution is supplied by psychoanalysis. Most of the Bohemians in the Village today are "under analysis" so that loyalty to the "school" to which their therapist belongs has replaced the commitment to artistic, political, and social credos of earlier decades. In addition, the inevitable expensiveness of psychoanalysis as well as its considerable duration forces patients to accept well-paying jobs which all too often contradict their artistic, intellectual, and personal values.[8]

While the prior stages of urbanization, industrialization, and bureaucratization in America called forth a sub-community—Bohemia—in which protest against the main drift was consolidated, there is no assurance that the present later stage of development will issue an equivalent. The need is obviously still with us, though this by no means guarantees that it will be met. Some students are encouraged by the emergence of the "hipster" on the borderland between the worlds of adoles-

[8] For comments on Greenwich Village during the thirties and forties, see the perceptive article by Milton Klonsky, "Greenwich Village: Decline and Fall" included in the volume *The Scene Before You*, edited by Chandler Brossard, Rinehart and Company, New York, 1955, pp. 16-28.

cence, jazz, homosexuality, and the slum. There is no evidence as yet that this social type even with his special uniform and language can evolve a set of workable countervalues or institutional patterns in any way comparable to those coming out of Greenwich Village in its prime. So far, the hipster looks more like an unfortunate symptom of, rather than a challenging response to, the pressures of modern life.

Yet it is still too soon to pronounce the patient dead. Our new conformity may yet evoke a new Bohemianism in the interstices of American city life. The danger is that sociologists will be so preoccupied searching for variants of the old models that they will miss the new.

DEEP SOUTH

DURING the late thirties two excellent studies of Southern towns were completed by American social scientists. The first, *Deep South* by Allison Davis, Burleigh Gardner, and Mary Gardner, is primarily a study of social organization and social control. The second, *Caste and Class in a Southern Town*, by John Dollard, supplements the sociological picture by examining the psychological effects that this social structure has on the people involved in it. Taken together, the two provide a good working conception of Southern small-town life during the thirties.

Both studies demonstrate the importance of regional patterns and problems. Yankee City, as its pseudonym implies, was part of the New England regional complex and Warner took this into account when analyzing its social structure. Thus, the upper-upper descendants of the ship owners formed a level of community stratification in this region not likely to occur elsewhere. In the Southern communities the old plantation-owning aristocracy occupies a comparable place, even to the disappearance of its economic base. Yet this similarity should not be pushed too far, since the existence of an upper-upper class of old families is not the main distinguishing feature of Southern sub-culture or social organization. This region is marked by a system of caste distinctions between White and Negro citizens which pervades every community so as to become the focal point of organization and control. Whereas Yankee City's ethnic patterns are fairly similar to those of any Northern, Midwestern, or Western town, all of these differ sharply from any caste-bound Southern community in this essential respect.

[153]

II. DEVELOPMENT

Recently, the legitimacy of this caste complex has been challenged by the Supreme Court with respect to segregated schooling, public recreational facilities and means of transportation. Though the issues involved are complex, these two community studies throw some light on the costs that this caste system entails for both groups. The Supreme Court used sociological and socio-psychological insights to show that "segregated but equal" facilities could never be fully equal, simply because Negroes were forced to use them while Whites did so voluntarily. At the heart of this contention lay the recognition that self-esteem is as important as more tangible prerogatives. Legal recognition of this fact was based on social-scientific research.

Since it is clearly impossible to deal with all aspects of Southern life reported in these community studies, emphasis will be placed on a few aspects of greatest theoretical importance for this book. First, the social structure described by Davis and the Gardners will be reviewed in an effort to identify points at which the processes of change might affect the caste system. Second, drawing mainly on Dollard, an interpretation of social disorganization in Southern communities will be advanced based on psychoanalytic as well as social structural observations.

Starting with *Deep South*, we find that one fact which emerges from the welter of data is the focus of all institutions in the community on the problem of keeping the Negro sub-community subordinate. This is a very specialized type of social control in that one sub-community controls the other to a point where members of the subordinate caste are denied their legal rights as American citizens. Southern community life is a gigantic conspiracy through which Negroes are kept from turning against their White masters by a network of control mechanisms ranging from deference patterns to lynch violence.

It is important to distinguish this particular instance of social control from the range of problems identified by the Chicago sociologists which arise from the necessities of coordinating urban sub-communities under the conditions of rapid urbanization. There the agencies representing community integrative interests, the courts, police, schools, and newspapers, while potentially susceptible to middle- and upper-class bias, are at least formally dedicated to protecting the integrity of individuals and sub-communities. In Old City, these same agencies are committed to defend the whole program of caste privileges and are manipulated by the Whites to guarantee that they do this.

As Whyte and Lynd show, the facts of power can transform supposedly neutral control agencies in the city into vehicles for realizing the interests of particular groups. The distinctive fact about the South is that this region sanctioned such transformation by declaring a whole segment of every community devoid of formal legal rights. This was and is still often done in defiance of the American Creed which guarantees such rights, a defiance that is legitimated, if at all, on the ground that Negroes are "inferior."

Obvious though it may seem, it is important to grasp the distinctiveness of this ideology. It sets up a system of social control which is overtly repressive and abandons even the illusion of democratic social processes. Negroes were defined as both biologically and socially inferior so that they had to be "cared for" by Whites who expected in return the privilege of exploiting them in various ways. Negroes were confined to menial or manual labor except within their own sub-community; they had to permit White men free access to their women; they were forced to submit to regular humiliation whenever they encountered Whites. All of these indignities had to be accepted "gracefully" since severe though not necessarily legal penalties loomed in the offing.

This is clearly a "deviant" system of social control. Once democratic values are accepted as normative, there are constant pressures toward disorganization inherent in the structure as long as Negroes are tempted to assert their legal or human rights while the reorganization mechanisms employed, whether technically legal or illegal, only reinforce the sense of deprivation and oppression. Given the context of our national value system, the social system of these Southern communities must be adjudged disorganized even though there is an appearance of "order" on the surface.

However, as these studies show, the inner stability of the system is not as complete as Southerners would like to believe. Even if the departures from American values can be rationalized by "racist" doctrine, the Negro sub-community does develop inner differentiations which hold promise for future changes. Part I of *Deep South* describes the social characteristics and clique patterns of each class within the two castes in full detail. The main point of interest here is that the Negro sub-community, though a subordinate caste without any channel for legitimate movement into the White caste, has developed its own internally differentiated class structure. Though upper-class Negroes seldom attain the economic status of middle-class Whites, they still can move upward within the limits of the Negro hierarchy. As they do so, Davis and the Gardners show how they take on middle-class White social patterns to the point where they over-conform, becoming more compulsively middle-class than the most mobile Whites.

This Negro mobility, despite the fact that it is confined to caste boundaries, violates the absolute superiority of Whites. As the authors point out, these mobile Negroes very often feel the full wrath of lower-class Whites who resent their capacity to attain economic status greater than their own. Yet it is the caste system itself which ensures this minimal Negro mobility by opening remunerative roles which Whites cannot or will not fill. These roles entail provision of special services like barbering, undertaking, and the like to the Negro sub-community as well as the central profession of the ministry, which constitutes the moral backbone of the Negro bourgeoisie. Even in this small town, then, the Negro sub-community has its own institutions and social identity.

Davis and the Gardners observe that there was no mechanism through which the Whites could ensure the absolute economic inferiority of all Negroes without violating two American principles—"the principle of the sacredness of private property, and the principle of free competition." If we take into consideration all of the other American values violated by the South, it is hard to see why these should have been constraining without examining the larger context. Here *Deep South* includes a good body of material that suggests possible points of weakness in the caste system. It will be impossible to review all of these, but a few having special relevance to the community processes studied earlier deserve mention.

There are some interesting differences between Old City and its surrounding rural areas which the authors comment on briefly: "The whole pattern of caste behavior and controls is also significantly affected by the differences between rural and urban situations. It is commonly recognized, for instance, that the rural Negro is more subordinated in his behavior than the urban Negro. In fact, the rural Negro frequently follows the formal pattern of caste etiquette exactly. . . . The urban Negroes often crowd whites in the stores, ignore them on the streets and sidewalks, and frequently are accused by whites of 'seeing how far they can go.' The whites are well aware of this difference in behavior and often speak with wistful approbation of the politeness and 'good manners' of the country Negroes, as compared to those in the city. The whites in a neighboring county frequently complained that the whites in Old City are too lenient with the Negroes and 'don't keep them in their places.'"[1]

This is explained as a direct result of the spread of urbani-

[1] Davis, Allison, Gardner, Burleigh B., and Gardner, Mary R., *Deep South*, The University of Chicago Press, Chicago, Illinois, 1941, p. 54.

zation even in this small town. In the country, most of the Whites are upper- and middle-class planters while the Negroes are mainly lower-class tenant farmers so that the degree of Negro subordination is maximized and regularized. In town the class positions of Negroes and Whites meeting in impersonal urban contexts are not likely to be nearly as uniform nor the expectations about deference as clear: "Thus, the upper-class and upper-middle-class whites frequently have little direct relations with any Negroes except house servants or, possibly, a few laborers. The lower-middle-class whites, however, have contact with all classes of Negroes through various occupational relationships in stores, as foremen or laborers, and so on. Urban lower-class whites often work beside Negroes in factories or on the river and may have Negro neighbors with whom they are friendly. This diversity of relations tends to prevent the formation of simple and uniform behavior patterns such as the rural Negroes show."[2] This must hold even more in really large urban centers where deference expectations become vague and impersonal class symbols like clothes provide a challenge to traditional color distinctions.

Business motives can also lead to the violation of caste patterns. White storekeepers dependent on Negro customers often drop derogatory terms of address as do salesmen with Negro clientele. Upper-class Negroes occasionally are treated with respect by lower- and middle-class Whites and the authors report a few instances of solidarity across caste lines between members of the upper-class and lower-class individuals. Yet these variations are still exceptions and most interaction between the castes is accompanied by ritual reminders of Negro subordination. Furthermore, any general improvement of Negro class standing is resented especially by downwardly mobile lower and lower-middle class Whites who are prone to use terror or violence to maintain their threatened "superiority."

There are signs of incipient development in Old City and

2 *Ibid.*, p. 55.

Old County that could affect the caste system profoundly. They emanate from beyond this community and therefore are only alluded to by the authors, especially in Chapters 20 and 21, dealing with the relations between caste, class, and the changing economy. It is evident that the old plantation system with its roots in the Old South is being seriously endangered by the industrialization of agriculture. The oppressive share-cropping or tenant system was already beginning to prove uneconomical and there is no reason to assume that this trend did not continue. Built around the exploitation of poor Negroes and poor Whites, its ability to compete with mechanized farming was under continual challenge.

Even Old County has been invaded by absentee owners and their managers respect local taboos no more than they have to: "These manufacturing firms hire labor as cheaply as they can get it, with the result that in industries where white workers have not been able to establish caste taboos, colored workers are employed to do much the same type of labor as whites. They may even be preferred to white workers, because they can be hired for a lower wage."[3] This has further ramifications: "Not only do these nonlocal industries tend to disrupt caste in labor, but they put into the hands of colored workers money which the local white storekeepers are extremely anxious to obtain. Since money has the highest value in the economic system, it causes white middle-class and lower-class storekeepers to wait upon colored patrons deferentially. It thereby increases the difficulties of adjusting caste, which seems to be essentially a structure of pastoral and agricultural societies, to manufacturing and commercial economies. This money economy likewise leads the group of entrepreneurs and middlemen to whom it has given rise—the most powerful group in the production of cotton because they control credit and therefore production—to be unmindful whether they buy cotton from a colored or white farmer, whether they sell food,

[3] *Ibid.*, p. 478.

automobiles, and clothes to one or the other, whether they allow nonlocal industries to subordinate the lower economic group of whites to the lower group of colored workers; they care principally about increasing their money. Even the local white farm owners prefer colored tenants to white, because they can obtain higher profits from the former."[4]

Again we see how the great solvents of industrialization and bureaucratization, filtered through the business community, were finally beginning to seep into this area. It was a slow process and its effects on the caste system were certainly not clear in the thirties. The depression seems to have accelerated economic pressures on employers to use Negroes wherever it was most profitable, even to the point of allowing them to fill somewhat more skilled jobs; but the basic social structure of Old City remained untouched.

It is important to reaffirm again the limitations of *Deep South* as a basis for assessing the structures and problems of contemporary Southern communities, especially those occurring in cities. Even in the thirties, Old City and Old County were slightly anachronistic. World War II shook up caste patterns by exposing both Negroes and Whites from this region to other types of inter-caste relations, and it is likely that the mass media have furthered this process over the last twenty years. We badly need new studies of Southern life, especially in light of the threats to the regional sub-culture arising from recent Supreme Court decisions. This historical event of the fifties has probably marked the present generation of Southerners as deeply as the depression did the people of Muncie.

However, before we leave the Southern community of the thirties, a glance at another study should deepen our understanding of the social processes underlying the caste system. Unfortunately, it will not make for greater optimism about the likelihood of the system's rapid disappearance. John Dollard's *Caste and Class in a Southern Town* shows that the forces

[4] *Ibid.*, p. 479.

sustaining caste operate at a deeper level than role analyses might suggest. One special value of this study is that it incorporates psychodynamic observations as they bear on social structure.

In order to make observations of this kind, Dollard had to use different methods of study. He never presents as detailed a picture of the social structure as we get from *Deep South*, but apparently his community conforms pretty much to that model. By setting his problem as the investigation of the motivational forces which keep Whites and Negroes conforming to a system which violates values they would otherwise affirm as Americans, Dollard focuses attention on aspects of social control hardly touched by Davis and the Gardners. He begins with the assumption that the caste system forces the Negro to suffer serious psychological blows as he plays the subordinate roles assigned to him. Dollard is not satisfied to take the Negro's apparent willingness to continue doing so for granted but rather wants to explore this through the use of depth psychology. Similarly, he is concerned with the psychic effects on the Whites of administering these blows to other human beings. In a chapter on his research biases he includes a field note describing his early reaction to Southerntown, a reaction, incidentally, which he later analyzed as due to his Northern origins: "These white people down here are very charming and really exert themselves to do friendly things once you are accepted, but they seem very much like the psychotics one sometimes meets in a mental hospital. They are sane and charming except on one point, and on this point they are quite unreliable. One has exactly the sense of a whole society with a psychotic spot, an irrational, heavily protected sore through which all manner of venomous hatreds and irrational lusts may pour, and—you are eternally striking against this spot."[5]

[5] Dollard, John, *Caste and Class in a Southern Town*, Doubleday Anchor Books, Doubleday & Company, Inc., Garden City, N.Y., 1957, p. 33. Original edition, Harper & Brothers, 1937.

II. DEVELOPMENT

Actually, the reaction was probably at least as much due to his psychoanalytic training, the same training that led him to: ". . . watch for the reservations with which people carry out formally defined social actions, the repression required by social conformity, and, in general, to see, behind the surface of a smooth social facade, the often unknown and usually unacknowledged emotional forces which drive and support social action."[6] In any event, Dollard's interest in depth psychological processes conditioned his whole approach to the community.

Where the Davises and the Gardners had the advantage of being able to make contact with the two main caste sub-communities, one couple being White and the other Negro, Dollard had to gather his information from a more limited perspective, that of a middle-class White male. He used his psychiatric training to overcome this limitation by having Southerntown Negroes come to his office for daily interviews in which their dreams and fantasies were freely expressed. The informants were motivated by a desire for self-insight as well as by Dollard's assurance that the research was designed to refute the myth that Negroes lack a complex mental life. Apparently, Dollard's depth data on Whites is based on ordinary social contacts viewed through everyday expressions of the unconscious, like jokes, slips, and non-verbal cues. In his modified therapeutic-investigative sessions, Dollard combined the depth interview and the life history—two techniques that permitted him to gain depth perspective as well as insight into the relation between milieu and personality. His methodological appendix on the life history as well as his book on the subject, *Criteria for the Life History*, are important contributions to the literature on techniques for studying communities.

Dollard's introduction of depth psychological data provides a perspective on the coercive aspects of Southern community life that the structural approach of Davis and the Gardners never captures. Though it is clear from *Deep South* that the

6 *Ibid.*, p. 38.

[162]

Negroes despise and fear intimidation by Whites, the ramifying effects of these fears and hatreds are never treated. Similarly, the emotional responses of the Whites to this latent
hostility is revealed in quotations but never analyzed in any detail. Thus the two studies actually supplement each other neatly.
They confirm each other's structural findings in all important
respects while at the same time each provides details in depth
that the other omits. The two studies show that authors with
different theoretical perspectives, but equally sound sociological training, are likely to perceive similar structural details even
though their interpretive foci are different.

However, it is important to recognize that Dollard gathered
a different *kind* of data. He had a more profound sense of the
"role masks" that social structures impose and their emotional
effects which become an important aspect of community life.
Thus, while both books describe the circumstances under which
White men have privileged access to Negro women, their observations as well as their interpretations of this pattern differ.
Davis and the Gardners deal with this in a section on endogamy and analyze structural interconnections between various
possibilities. Thus, the taboo against relations between Negro
men and White women is seen as a way of protecting endogamy, so that its violation becomes the main occasion for violence; while the reverse relationships—that between White
men and Negro women—is encouraged as long as it is not
legalized or flaunted. Negroes usually insist that relations between White men and their Negro mistresses are "family" relationships and some Whites treat these relationships in quasi-
familial terms. Negro women benefit from a stable relation
with a White man and sometimes choose this in favor of legal
marriage to another Negro. This places severe strain on Negro
males:

"It appears, then, not only that the Negro man is subordinated in all his relations to the whites but that his subordinate
role weakens his relations with women of his own group. If

he marries, his wife may at least sometimes compare him with that potential ideal, a white lover. If he seeks after the more attractive girls, or those educated or better dressed, he must compete with the white man who can offer money, prestige, and security. The hopelessness of the situation was shown in a story told by an upper-class Negro girl about a Negro boy who approaches her in behalf of his White employer. When she berated him for his action, he replied: 'I'd like you myself, yes, indeed, I'd like you fo' myself, but I works for him.' "[7]

Other aspects of the situation are examined, including the situation of the "mixed breed" children and the possibilities of "passing" which appear less likely in a small community like Old City, where people know each other, than in larger cities or outside the South.

This description of caste endogamy with its picture of systematic inter-caste sexual relations is quite similar to that reported by Dollard. The functional analysis of the taboo on White female - Negro male sexual relations as defending endogamy and caste is far clearer than the analysis of permitted White male - Negro female intercourse, which on the face of it would seem likely to disrupt caste functioning because the ties involved in the permanent quasi-familial relationships cut across established caste lines. Davis and the Gardners never interpret this beyond the view expressed above that it is another aspect of Negro subordination and therefore congruent with caste principles.

In his chapter "The Sexual Gain" Dollard deals with the same problem of inter-caste sexual relations, but his interpretation employs different concepts to illuminate facets untouched by the sociologists. He begins by defining the access that White men have to Negro women as a "gain" for both, though it soon becomes clear that he is using the term in the sense of secondary "neurotic" gain as it is customarily employed by psychoanalysts, since the consequences of this differential

[7] Davis, et al., *op.cit.*, p. 38.

[164]

access are of dubious real value to either group. To establish the dubious character of these "gains," he points to the contrasting imagery among Southern White males in which White women are seen as pure and untouchable, while Negro women appear to be seductive and available. He then links this split image to the problem of social control of sexual behavior which is a central aspect of functioning in any community, but one rarely dealt with in explicit terms. These Southern communities force the dichotomy into focus simply because it is so problematic in all of them. The split image, establishing a scheme in which goodness is equated with chastity and badness with sexuality, is certainly not confined to Southerners. Analysts of the mass media such as Wolfenstein and Leites in their book *Movies* find this theme recurrent, in the form of good-bad heroines who combine apparent sexual looseness with real chastity. It is only in the South, however, that the split has been institutionalized so that one component is fixated on Negro females and the other on White females with the need for combining the two in the context of a mature relationship thereby obviated, or perhaps avoided. Another aspect of the stereotype is the White assumption that Negro men are also more virile and there are a number of reinforcing images to this effect corresponding to those about Negro women. This gives added vigor to the White males' responses when signs of Negro male interest in White females appear. Dollard shows that White women are themselves often preoccupied with this presumed super-sexuality of Negro males to the point where unconscious seductive behavior is likely to have been involved in many so-called "attacks."

Dollard presents a very subtle picture of the web of unconscious attractions and fears that underlie sexual relations between the castes and within each caste. His presentation documents the bias mentioned earlier regarding the psychic "sickness" of the South by exposing the warped sexual identities forced on Whites and Negroes alike by the caste system.

White females are almost forced into frigidity by their husbands, who get "sexual" satisfaction from Negro females without having to acknowledge their social existence. Negro males are forced to watch their females being exploited by White males without intervening and thereby suffer blows to their own masculine self-esteem while the Negro female cannot heal this wound since she is the unwilling or perhaps even willing participant in the process. There is more than a bit of irony in Dollard's final comment:

"A qualification is necessary concerning the sexual gain of middle-class whites. The fact that the white men have access to Negro as well as white women and fight for the situation which allows them this privilege does not mean that they can enjoy these sexual contacts to the utmost; nor does it mean, we repeat, that all white men who have such privileges make use of them in fact. Many times people are cheated of the enjoyment of hard-won gains by the development of conscience. Impulse, once controlled, may be controlled too well. Marriage, for example, does not always mean release of the sex impulse. The middle-class white men in this town have superior prerogatives. The question remains as to whether the strain of becoming a middle-class individual does not frequently damage ability to enjoy these prerogatives. The lower-class Negroes, though actually more limited in sexual choices, may have much freer psychic capacity to utilize the women available."[8]

However it may turn out, the fact remains that no one in this caste-ridden community is encouraged to acknowledge members of the opposite sex as full physical, social, and sexual personalities nor are they encouraged to acknowledge themselves as such.

Another "gain" of the middle-class Whites is automatic prestige: the mere fact of being white-skinned entitles the holder to deference from any black-skinned persons who come into his presence. Dollard describes the obvious "narcissistic wound"

[8] Dollard, op.cit., p. 172.

that this entails for the Negro, who must despise his own skin color, but he also suggests that acceptance of this self-humiliation is impossible without deep hatred against those who impose it. This hatred cannot be expressed openly but most Whites are aware of it with varying degrees of consciousness, as they are aware that their whole network of exploitative "gains" is deeply resented by the Negroes who endure them: "It struck me repeatedly that the deference of Negroes, in addition to being pleasant, has the function of allaying anxiety among white people. By being deferential the Negro proves, in addition, that he is not hostile. Whites are not satisfied if Negroes are cool, reserved, and self-possessed though polite; they must be more than polite; they must be actively obliging and submissive. It would seem that there is much fear behind the demands of white people for submissiveness on the part of Negroes."[9] In a democratic society, systematic attacks on the self-esteem of others leave the attacker guilty and anxious even if these actions are "legitimated" by caste.

In his exploration of "the emotional forces which drive and support social action," Dollard devotes four chapters to describing and analyzing the patterns of aggression management adopted by each caste. This is at least as central as the channeling of sexual impulses and its "ordering" by the social structure raises closely related problems. In the face of the "gains" that the Whites extract from them, Southerntown Negroes can:

"1. Become overtly aggressive against the white caste; this they have done, though infrequently and unsuccessfully in the past.

"2. Suppress their aggression in the face of the gains and supplant it with passive accommodative attitudes. This was the slavery solution and it still exists under the caste system.

"3. Turn aggression from the white caste to individuals within their own group. This has been done to some extent and is a feature of present-day Negro life.

[9] *Ibid.*, p. 185.

"4. Give up the competition for white-caste values and accept other forms of gratification than those secured by the whites. This the lower-class Negroes have done.

"5. Compete for the values of white society, raise their class position within the Negro caste and manage aggression partly by expressing dominance within their own group and partly by sheer suppression of the impulse as individuals. This is the solution characteristic of the Negro middle-class."[10]

Dollard discusses each of these possibilities in detail and is thereby led to many acute observations about Southern community patterns which slipped through the coarser institutional net of Davis and the Gardners. His psychodynamic focus permits interpretation of underlying motives and ambivalences which facilitate understanding of many structural patterns that might otherwise appear utterly irrational.

One of the chapters that best demonstrates the applicability of psychodynamic interpretations to social phenomena is that dealing with "defensive beliefs." Here Dollard analyzes various components of the stereotypes that Whites hold about Negroes to show specifically how each element helps to rationalize White aggression and exploitation by: ". . . concealing the disparity between social justice according to our constitutional ideal and the actual caste treatment of the Negro." He goes on to observe that: "Defensive beliefs in a society, like rationalizations of the individual, make possible the avoidance of the actual situation; they tend to eliminate the problematic, offensive, inconsistent, or hostile facts of life. One clings to the defense and avoids the traumatic perception of social reality."[11] Several specific beliefs are taken up and the reality distortions involved explained in terms of the socio-psychological defenses invoked. Thus, the assumption of Negro "animality" is seen to justify White prerogatives which involve exposing Negroes to continual humiliation by assuming them

10 *Ibid.*, p. 253.
11 *Ibid.*, p. 367.

to be creatures who, like "animals," do not have feelings that can be hurt.

Dollard offers some far more complex explanations such as the tour-de-force that follows wherein he interprets the multiple functions of the White belief that Negro men invariably desire White women. His reversal of the Freudian assumption by suggesting that this belief masks a deeper conviction that Negroes really want equal status rights is a neat twist. The entire passage shows the intricate interweaving of status defenses with sexuality and aggression:

"Behind the actual or posited desires of Negro men for white women is seen also the status motive, the wish to advance in social rank, to be as good as anyone else, and to have available whatever anyone else has, that is, free sexual access to women, within the mores and without undemocratic distinctions. The race-conscious white man has a proprietary interest in every white woman; he also acts as if an attack upon or an approach to the white woman were an attack on himself. The horror generated by the rape idea resembles closely the dread which individuals show concerning brutal and sadistic operations on their own bodies. Perhaps the path of this unconscious process is something like the following: the question is asked, 'Would you want your sister to marry a nigger?' This amounts to asking whether an individual would defend his sister against an attack by a Negro in case of a pitched battle; and the further question is raised, 'What harm could come to you if you did thus defend her?' It is a matter of fact that the approaches of white men to Negro women are interpreted as assaults on the dignity of the Negro man, and no fight occurs because the Negro is categorically prohibited from fighting at all. Out of such considerations arises the conviction of guilt of Negroes and the assumption of the wish to attack white women. This belief has the great value of concealing the actual sexual gains of the white caste which are still accessible to those who are disposed to take advantage of them; that this gain is real is shown by the number of mixed-blood Negroes. Behind the fears of retaliation and revolt lie the social facts of exploitation of Negroes; fear of revolt conceals the true state of affairs."[12]

There are so many genuine insights in this passage that commenting on them all would take far more space than the original. It is included here as a model of social psychiatric interpretation linking emotional processes to community structures and functioning. Dollard's conception of levels of interpretation in which group stereotypes can defend status gains and aggressions might well apply to inter-group relations in quite different communities.

There is a great deal more worth analyzing in *Caste and Class in a Southern Town*. Dollard reports on "gains" of the lower-class Negroes and interprets the symbolic meaning of caste and race prejudice. It is true that lower-class Negroes do achieve considerable directness of impulse expression, though expressions of potentially dangerous sexual and aggressive impulses are carefully confined to their own group. Because of this emotional directness, they do not require sublimated or substitute gratifications and even their religion follows this pattern. Dollard had noted the role of religion in promoting caste acceptance first, through ideological rationalization that suffering on earth (caste subordination) would be rewarded in heaven and, second, by providing an important and immediate source of compensatory emotional satisfaction through its highly expressive ritual.

Lower-class Negroes become irresponsible or "carefree" mainly because they have no motivation to accept responsibilities. Their so-called "shiftlessness" becomes the only workable response to a desperately depriving environment. Middle-class Negroes, too, have a constant struggle to keep from deteriorating in a similar fashion when they compare their opportunities with those of middle-class Whites. Middle-class discipline is painful not only to Whites but even more so,

[12] *Ibid.*, pp. 382-383.

because it is less rewarding, to Negroes; and as a result both middle-class groups tend to idealize the "freedom" of the lower-class Negroes and end up by simultaneously envying and despising him.

Moving to a deeper level of analysis, Dollard suggests that the underlying theme of caste subordination is Negro childishness. Whiteness becomes the symbol of personal maturity and dignity. This is taken over by the Negro sub-community in which lightness of skin color is prized and darkness condemned under most circumstances. However, this quasi-parental role assumed by Whites is quite unlike real parenthood because it aims at keeping the Negro permanently immature, at denying him the right to full growth, which is the true object of ordinary child-rearing.

This whole social structure, which is based on one sub-community exploiting another, invariably affects the White exploiters as well as the exploited Negroes. Dollard ends his book by pointing to the psychological mechanisms involved in race prejudice: "Our conception works with three key concepts. First a generalized or 'free-floating' aggression which is derived from reactions to frustration and suppression within the 'we-group.' It can be thought of as a tendency to kick, hit, scorn, or derogate someone or something if one could only find out what. A second necessity is that of a permissive social pattern. This must exist in order to lift the in-group taboos on hostility. The permissive pattern isolates a group within the society which may be disliked. Usually it is a defenseless group. In the South the caste role of the Negro is the pattern which permits white people systematically to derogate him. The permissive pattern often comes down from earlier days, although it can also be invented on the spot or by analogy with a group against which one is already prejudiced. The third essential in race prejudice is that the object must be uniformly identifiable. We have to be able to recognize those whom we may dislike. This stipulation is met by the physical or cultural marks which make the object of race prejudice 'visible.' In our sense race prejudice is always irrational. . . ."[13]

The applicability of these observations beyond Southern-town is unquestionable. Their purpose here is to show that White Southerners are displacing aggression so that attention is diverted from real inadequacies in the communities in which they live onto the Negroes. Their caste superiority gives Whites a minimum of self-esteem but, as Dollard has shown, the price paid is utterly distorted perception of Negro motives, feelings, and qualities. Maintaining these distortions entails blinding themselves both to their own cruelty and to the suffering as well as the aggression it engenders. It reaches back into their family lives by rendering the men dissatisfied with their own women and the women similarly dissatisfied but with no outlet for their sexuality. Violence becomes a regular aspect of Southern life, taking its toll of White sensibilities as well as Negro lives. Furthermore, no amount of defensive rationalization can protect White Southerners from some guilt over the violations of basic American values which the caste system entails. This guilt may only lead them to further aggressions against the Negroes but violence cannot restore their sense of congruence with the "national identity" any more than can frequent displays of super-patriotism. Perhaps the persisting myth of the Confederacy is some compensation, but, as centralization increases, the impoverishment of this myth, as measured by the South's inadequate social and economic standards, is increasingly evident.

Later studies like Hunter's *Community Power Structures*, which contains a fascinating chapter on the power position of the Negro sub-community in a large Southern city, suggest that inner pressures for change in the caste system coming from the Negro community itself will eventually link with outer pressures emanating from the Federal Government and

[13] *Ibid.*, pp. 445-446.

Northern groups, both Negro and White, to seriously modify caste requirements. The analysis of these social systems shows that they are rigidly integrated and suggests that persons seeking change will have to protect themselves and members of the weaker caste against the emotional and potential physical violence which is likely to erupt. Much, however, has happened since the thirties, and these studies can certainly no longer be taken as representative of present-day community structures even in rural areas and certainly not of those now dominating Southern city life.[14]

Deep South adds a new range of variation—regional complexes—to the theory of communities. Furthermore, it describes a kind of regional community integration around caste which creates overwhelming problems of social control not found to the same degree in other regions, nor solved in the same way in any of them. The fact that this system exists in a democratic national context renders it unusually vulnerable, and even its firmest adherents have special problems because of this. Davis and the Gardners do show some of the points at which urbanization, industrialization, and bureaucratization are likely to transform, if not eliminate, many of the distinctive regional characteristics even including caste.

Caste and Class in a Southern Town makes another kind of contribution to the theory of communities. It demonstrates the importance of analyzing the covert emotional life of the community and offers a highly sophisticated model for doing so. Dollard does not fall into any oversimplified theories about "basic" or "modal" or even "status" personality. Instead he uses his knowledge of psychoanalytic processes to show how persons with differing characters are forced to adapt to the

[14] Were it not for the fact that such a discussion would have expanded this chapter into a separate monograph, several excellent Southern studies, including Hunter's book, *The Blackways of Kent*, by Hylan Lewis, *The Millways of Kent*, by Ken Morland, and *Plantation County*, by Morton Rubin, would have been used to bring the analysis of Southern communities up to date.

role-playing alternatives confronting them. The range of alternatives is itself sociologically determined and depends on their position in the class-caste configuration as well as on their age and sex. Deep-lying latent points of tension are thus identified, as are control mechanisms inaccessible to analyses concerned only with social structure. Dollard's study also deepens our concept of community disorganization in that it shows how an overtly orderly community can exact tremendous emotional penalties from its members. Actually, Southerntown dehumanizes both Whites and Negroes so that members of the two castes respond to each other categorically instead of concretely. The further implication that they also are forced to respond to each other stereotypically becomes even more poignant when the last possible consequence—that they also respond to themselves in this fashion—is made explicit.

WORLD WAR II AND MILITARY COMMUNITIES

THERE are no community studies which do for the years of World War II what the Lynds' *Middletown in Transition* did for the depression era. But there are a number of partial studies that throw light on different aspects of community life during this interval without quite getting at the interplay of basic social processes as have the best community studies. Part of the problem lies in the fact that many sociologists were themselves engaged in military service of one kind or another so that their availability for research of this kind was limited, while those at home often had staggering teaching loads. Luckily, however, sociologists and social psychologists working in and with the Army did produce a worthwhile series on Army life. By broadening our conception of a community study, we can examine the special military communities in which most adult males found themselves.

Other studies undertaken during the war are all severely limited in scope and theoretical relevance. The best of them, like *The Social History of a War-Boom Community* by Robert Havighurst and H. Gerthon Morgan or Alexander Leighton's study of a Japanese relocation center, *The Governing of Men*, deal with such special kinds of war-molded communities that their usefulness for incorporation into this kind of a general theory is limited. Seneca, Illinois, the "war-boom community," expanded five-fold, underwent serious institutional change, and then, at the war's end, returned pretty much to the original state of affairs. The detailed discussion of accommodation and conflict between Oldtimers and Newcomers could probably be generalized for other boom-towns but its relations to the more fundamental processes accompanying urbanization, industrialization, and bureaucratization remain unexplored. Similarly, Leighton's report on Poston would apply to the problems of organization and disorganization in a relocation setting but its wider relevance is uncertain.

Since the theory being developed here aims at delineating the historical contexts in which community development took place over the past fifty years, the omission of the war period because of lack of an entirely satisfactory study would leave a serious gap. This justifies the introduction of alternative sources of data if they can be found. In this instance, there is one such source. The first two volumes of *The American Soldier* by Samuel Stouffer, et al., contain a broad range of material on the experiences of Americans in the Army during World War II. Because of the project's tremendous scope, only a small portion of this material can be introduced here. But that portion does help to illuminate important community experiences and provides a sort of bridge from community experiences during the depression as reported by Lynd to the epoch of the suburb that constitutes the subject matter of the next chapter.

Actually, the findings in these volumes were obtained by techniques quite different from those employed in the other studies we have discussed. Interviews and observation were only supplementary; the bulk of the data was gathered through questionnaires which were then subjected to statistical analysis. Most of the specific studies undertaken were directed at answering questions of special concern to military sponsors; their relevance for more theoretical problems has been demonstrated, however, by the reorganization of primary data in the volumes themselves as well as by later reanalyses like those contained in *Continuities in Social Research*, edited by Robert Merton and Paul Lazarsfeld. This chapter is another such reanalysis, designed to show some of the contributions of this material to a theory of communities.

If we recall, then, that the soldiers studied in these volumes were the same men whose responses to the depression Lynd described earlier, Hans Speier's comments on their attitude toward the war seem less surprising: "The data in *The American Soldier* on the personal commitment and on the orientation of soldiers toward the war present a gloomy picture. While some of the individual parts of the picture almost certainly result from the use of methods which do not permit any delicate probing into motivations and convictions, the composite picture leaves no doubt that the American soldier had neither any strong beliefs about national war aims nor a highly developed sense of personal commitment to the war effort. He did not think much about the meaning of the war as a whole and displayed a tendency to 'accept momentarily any plausibly worded interpretation of the war.' "[1]

After summarizing the attitude study findings to show that American soldiers had a very limited personal commitment to the war enterprise and low insight into our war aims, he remarks: "If the United States had lost the war, there can be little doubt that these findings would have been drawn upon to explain the defeat. It is indeed exceedingly difficult to understand, on the basis of the research here reviewed, why the American armed forces fought as well as they did."[2] Yet this syndrome is not surprising if one thinks back to the attitudes toward large-scale structures—political, economic (and now military)—which such earlier students of the community as Lynd and Warner found during the depression. In a way, the American soldier was simply transferring his essentially detached, though dependent, attitude toward the Federal government and his corporate employers to the war and the Army. Continuity is more evident than discontinuity.

It is now necessary to raise some fundamental questions

---

[1] Speier, Hans in *Continuities in Social Research*, edited by Robert K. Merton and Paul F. Lazarsfeld, The Free Press, Glencoe, Illinois, 1950, p. 116.
[2] *Ibid.*, p. 117.

about military motivation. Due to their lack of historical perspective, the authors of *The American Soldier* painted a collective portrait that often appears paradoxical at precisely the points where paradox is unnecessary. Speier calls attention to one such paradox: how could an Army whose soldiers had so little dedication to or understanding of the cause for which they fought fight so successfully? The authors of *The American Soldier* raise this question most directly when they note that the lack of patriotic incentive appears even stranger in light of the dangers of combat. Motivation was supposed to have been supplied by the network of personal ties in the small combat groups to which these soldiers belonged. Rather than looking to patriotism or other "lofty" motives, the "new" social psychologists turn to the small group as the key factor.

Discovery of small groups as the key motivating factor for filling otherwise unpleasant social roles was a convergent finding made by many American social scientists during the late thirties and forties. It turned up as the main variable in such places as factories, military units, neighborhood gangs, and almost every place else where there was more than one person. By now it has become nearly as ubiquitous among American social scientists as sex for psychoanalysts. Without claiming to have done the spadework to put this convergence in true historical perspective, we might suggest the hypothesis that the emphasis by sociologists during the late thirties and forties on small groups was itself a response to the deterioration of large-scale orientations and commitments. Since they were caught up in the social processes they studied, it was logical for them to interpret these processes in terms of a "vocabulary of motives" that made sense in their own experience. Later on, the willingness of the authors of *The American Soldier* to explain military motivation in primary group terms rather than to view this kind of motivation as itself a product of the deterioration of ideological commitment is another manifestation of the same thing. By 1950 the authors have lost contact with the historical process to such an extent that they occasionally assume they have discovered in the primary group the essential basis for motivation to fill military roles overlooked by previous students of army life.

In any event, one of the main themes running through *The American Soldier* is the immense importance of primary groups for military motivation. No intelligent observer would categorically deny this, yet it is something that must be studied in historical perspective if its true significance is to be assessed. Dollard's study of Spanish Civil War veterans, *Fear in Battle*, shows that they were far more likely to say they were motivated by ideological commitment than by primary group ties and there is reason to suspect that the place of such ties in the balance of military motivation throughout history has not been as important as it apparently became during the last war.

During the thirties, our Army had never grown larger than 200,000 men. Even depression unemployment had not created sufficient pressure to drive any significant proportion of the male population into this low-status way of life. With Selective Service in 1940, men were recruited on a compulsory basis, but it was really not until after Pearl Harbor that public opinion galvanized to support mobilization. Men entering the service during these early years did not have very high respect for Regular Army soldiers and the authoritarian caste system to which they were rapidly subjected did little to heighten it.

Chapter Two of Volume 1 of *The American Soldier*, "The Old Army and the New," contains a great deal of information about the responses of draftees at the time of Pearl Harbor to the Army they had been forced to enter. This was an organization in which membership was obligatory, with no prerogatives regarding the termination of duties or even channels for complaining about orders considered unjust. These draftees had to accept the fact that officers could give them arbitrary orders which had to be complied with, while at the same time these officers enjoyed privileges, which draftees were categorically denied. Thus, the Army's "caste" system became a major focus of resentment, especially against the background of involuntary servitude.

Even though Pearl Harbor helped to legitimate the need for an army, it did not completely legitimate each individual's service. The well-established American tradition of equal "sacrifice" remained operative and it was this which gave rise to the important concept of "relative deprivation" that constituted one of the main conceptual innovations of the two volumes:

"Becoming a soldier meant to many men a very real deprivation. But the felt sacrifice was greater for some than for others, *depending on their standards of comparison.*

"Take one of the clearest examples—marital condition. The drafted married man, and especially the father, was making the same sacrifices as others plus the additional one of leaving his family behind. This was officially recognized by draft boards and eventually by the point system in the Army which gave demobilization credit for fatherhood. Reluctance of married men to leave their families would have been reinforced in many instances by extremely reluctant wives whose pressures on the husband to seek deferment were not always easy to resist. A further element must have been important psychologically to those married men who were drafted. The very fact that draft boards were more liberal with married than with single men provided numerous examples to the drafted married man of others in his shoes who got relatively better breaks than he did. Comparing himself with his unmarried associates in the Army, he could feel that induction demanded greater sacrifice from him than from them; and comparing himself with his married civilian friends he could feel that he had been called on for sacrifices which they were escaping altogether. Hence the married man, on the average, was more likely than others to come into the Army with reluctance and, possibly, a sense of injustice.

"Or take age. Compared with younger men—apart now from marital condition—the older man had at least three stronger grounds for feeling relatively greater deprivation. One had to do with his job—he was likely to be giving up more than, say, a boy just out of high school. Until the defense boom started wheels turning, many men in their late twenties and early thirties had never known steady employment at high wages. Just as they began to taste the joys of a fat pay check, the draft caught up with them. Or else they had been struggling and sacrificing over a period of years to build up a business or profession. The war stopped that. Second, the older men, in all probability, had more physical defects on the average than younger men. These defects, though not severe enough to satisfy the draft board or induction station doctors that they justified deferment, nevertheless could provide a good rationalization for the soldier trying to defend his sense of injustice about being drafted. Both of these factors, job and health, would be aggravated in that a larger proportion of older men than of younger men got deferment in the draft on these grounds—thus providing the older soldiers, like the married soldiers, with ready-made examples of men with comparable backgrounds who were experiencing less deprivation. Third, on the average, older men—particularly those over thirty—would be more likely than youngsters to have a dependent or semi-dependent father or mother—and if, in spite of this fact, the man was drafted he had further grounds for a sense of injustice."[3]

These passages show one analytic model used by the authors of these volumes to good advantage. A set of empirical findings—that married men and older men resent being drafted more than their unmarried and younger counterparts—is interpreted using the social-psychological conception of relative deprivation to identify several reference groups in terms of

[3] Stouffer, Samuel A., et al., *The American Soldier*, Vol. i, Princeton University Press, Princeton, N.J., 1949, pp. 125-126.

which persons in these categories were able to feel "relatively deprived." Historical continuity is seen in the recognition of the depression background. On a broader plane, the wide spread sense of "relative deprivation" itself derives at least in part from American emphasis on equality of sacrifice and shows the secular terms in which we accepted military service.

As may have been expected, Americans approached military service with the democratic expectation that everyone would be sharing equally in the burdens and rewards involved. DeTocqueville commented on this aspect of democratic armies:

". . . Men living in democratic times seldom choose a military life. Democratic nations are therefore soon led to give up the system of voluntary recruiting for that of compulsory enlistment. . . .

"When military service is compulsory, the burden is indiscriminately and equally borne by the whole community. This is another necessary consequence of the social condition of these nations and of their notions. The government may do almost whatever it pleases, provided it appeals to the whole community at once; it is the unequal distribution of the weight, not the weight itself, that commonly occasions resistance."[4]

This is a slightly more general statement about the origins of this sense of relative deprivation. It places the phenomenon in a larger historical context, that of democratic armies, on the one hand, and points to the specific value, "equality of sacrifice," which makes it operative on the other.

One of the most important empirical facts emerging from *The American Soldier* is that the main dimension of relative deprivation in the American Army during World War II was the feeling on the part of enlisted men that officers systematically and unjustly received special privileges as a result of the caste system. Attacks on this system of special privilege

[4] De Tocqueville, Alexis *Democracy in America*, Vintage Paperback Edition, Vintage Books, New York, 1954, Vol. II, p. 286.

[182]

were a major aspect of everyday life among enlisted men, and there is some evidence to show that it became a scapegoat for all frustrations occasioned by Army life. Some comments from enlisted men in the Persian Gulf command said to be fairly typical of feelings expressed in other theaters illustrate some dimensions of the problem:

"The distinction made between officers and men is so great that it spoils any attempt to raise our morale by movies and footballs. All we ask is to be treated like Americans once again. No 'out of bounds,' no different mess rations, and no treating us like children."

"Why must the enlisted men be confined to camp as though he were in a concentration camp, when the officers can go where they dam please. The officers go to town, the officers get the few available women; there are several social affairs given from time to time for officers but nothing for the enlisted man unless it be an exciting bingo party. My pet peeve— to see a commissioned officer out with a girl flaunt her in front of enlisted men, who cannot go out with nurses."

"The officers are getting American whiskey and we are not. I do not think it's fair."

"Too many officers have that superior feeling toward their men. Treat them as if they were way below them. Many of the men have just as good an education, if not better, than many officers and also have come from just as good families. What's the matter with us enlisted men, are we dogs?"[5]

These comments and many more suggest something of the bases for resentment against officers. One important component was clearly the differential access to consumer privileges as well as to women which this caste system imposed. This was more marked in overseas garrison situations than in training or combat.

A statistical comment on attitudes toward officers showed that three times as many men had unfavorable attitudes as

[5] Stouffer, *op.cit.*, Vol. I, pp. 371-372.

had favorable attitudes; expressed differently, 40 per cent were definitely hostile; 50 per cent were relatively indifferent, and 10 per cent were favorable. These findings were based on attitude studies scaled in terms of intensity scores so that many of the "indifferents" had some unfavorable attitudes. The questions dealt with practically all aspects of officer-enlisted man relations and general feelings about officers. The results are reported in Volume i of *The American Soldier*, pages 375-379.

Hans Speier's perceptive interpretation of attitude data in these volumes, using Mannheim's theory of social perspectivism, points to the serious disagreement between officers and enlisted men over caste privileges as against their general agreement on more neutral matters. He uses this carefully elaborated finding to demonstrate the importance of specifying the kinds of knowledge that are affected by the social perspectives of the holders and suggests that matters linked to status threats and defenses are more likely to show "perspectivistic" distortions. Thus the fact that 67 per cent of the officers agree that they "deserve special privileges because of their greater responsibility" while only 23 per cent of the enlisted men feel this to be the case ensures the likelihood that these special privileges will become a source of contention between the two groups.

Here the question might be raised: "How is it that officers coming from the same cultural system with its emphasis on equality of sacrifice are able to accept their unequal privileges?" The first point is that the 67 per cent acceptance suggests that there are a good number who do not quite accept these privileges; but, even so, the large majority still do. Luckily, the authors did include some material on socialization in Officer Candidate Schools which shows how officers learn to deal with this problem: "A case might be made for the hypothesis that the experiences of a candidate in Officer Candidate School not only did not involve much explicit in-

[184]

struction in handling enlisted men, but actually contributed indirectly to making it harder for the new young officer to see the enlisted viewpoint. The Officer Candidate School could be conceived of as an ordeal, with some functions not dissimilar psychologically—in their emphasis on hazing and attention to minutiae—to those of ordeals involved in a college fraternity initiation."[6]

The details of this shift in orientation on becoming an officer are beautifully described and analyzed in a participant-observation report by M. Brewster Smith of which only a few points can be quoted. Smith observes that would-be officers are subjected to an ordeal of "close discipline" with the constant threat of "washing out" to keep them in line and the promise of release if they last to become upperclassmen. Becoming identified with the source of this discipline is an adjustive response. Progressive identification with the officer role is achieved by allowing officer trainees to assume upper-class prerogatives gradually during the latter stages of OCS. Aggression repressed earlier is allowed gradual release in such a fashion that aggressive enforcement of disciplinary patterns imposed on enlisted men becomes part of the restored ego equipment of the candidate. The upperclass man in OCS "pays back" the classes following him for the indignities he suffered earlier when he was in their position. Later, he treats enlisted men in the same fashion: "The new officer, somewhat insecure in his role and perhaps a little guilty at his favored status over his previous enlisted confreres, reactively asserts his status, and finds in the OCS ordeal a justification for his new prerogatives; he *earned* them. The means whereby he earned them come to have special value for him. He puts a high value on official 'GI' ways of doing things, and rationalizes that what was good for him must be good for those under his command."[7] As it happened, the enlisted men neither knew nor cared about this "ordeal" so that they tended

[6] *Ibid.*, p. 389.   [7] *Ibid.*, p. 390.

to interpret the "overbearing" behavior of "90 day wonders" with less understanding and more indignation than the Army or the "wonders" would have liked.

Army newspapers run by enlisted men often reflected this hostility toward officers and the caste system. *Stars and Stripes* had a column called "B-Bag" which was filled with comments on this topic. Much of the humor of George Baker's wonderful cartoon character "Sad Sack" depends on Chaplinesque irony about the Army's caste system. Perhaps the best record of this kind is found in Bill Mauldin's brilliant and occasionally savage cartoons, now collected in the volume *Up Front* though originally appearing regularly in Army newspapers all over the world. These cartoons capture more of the mood of American soldiers than any of the novels or other art works that came out of the war. Two savage commentaries on the caste system come to mind: the first shows two officers looking at a spectacular sunset while one turns to the other and comments: "Beautiful view. Is there one for the enlisted men?" In the other, an obviously battle weary GI is seen lying in a hospital bed surrounded by two Medical Corps captains and a major, one of whom says thoughtfully to the others: "I think he should at least try to lie at attention." Mauldin did other cartoons in the same vein, contrasting human realities with the authoritarian rigidities of caste thinking.

After all this manifest resentment, the question that arises is, "Why did the Army maintain and defend the caste system?" There are several levels on which this question can be discussed. In broadest historical terms, military forces have always needed some kind of impersonal authority system simply because the requirements of most military operations revolve around closely coordinated movements of large groups of men in such a fashion that policymakers at the top can be certain that their commands will be carried out by the officers and men under them. Of course, this is less true in other special circumstances, like guerrilla armies, but these exceptions are not significant for this general survey.

On a somewhat lower level of generality, caste systems and discipline would seem to be particularly important in a democratic army simply because the soldiers in it have been accustomed in civilian life to individualized patterns of behavior and must therefore be trained or broken to an entirely new pattern. It would be easy to exaggerate this point, since, as we know, factory work and many other occupations leave only a small margin for individual discretion; yet the pattern of "free choice" does remain in such areas as leisure and family life. To inculcate a pattern of instantaneous obedience under these circumstances is probably far more difficult than in an authoritarian milieu. Thus, the American Army in World War II was confronted with the problem of transforming several million "individualists" with widely varying backgrounds into interchangeable troop reservoirs so that performance within minimum limits could be predicted on a sufficiently safe basis to permit military operations. Here, the special problem of developing generalized attitudes on the part of enlisted men toward authority which would lead them to follow the commands of any superior officer was unusually acute because of the American expectation that leaders have to "prove themselves" before their authority is legitimate. Given the fact that officers more than anyone else are likely to be killed in combat, maintenance of a chain of command presupposes acceptance of the authority of the next in command, no matter who he may be. In other words, the problem of succession so aggravated in combat makes it even more essential that military discipline be imposed on a rigorous basis.

There is still another kind of problem that one suspects is more likely to erupt in democratic armies, especially in the American Army. This problem has been the theme of innumerable Hollywood war movies, and it must conceal some kind of truth about our life, even though that may not lie in its manifest message. This is the situation in which an apparently "hard-hearted" company commander (or flight commander, ship's captain, or first sergeant—it does not seem to include corporals) despised by his men as a cruel sadist is shown during the torturous development of the plot to be really deeply devoted to his men—so much so that sending them into combat and possible death evokes tenderness. But he knows that befriending them might incline him toward partiality or, worse yet, toward protecting his outfit at the Army's expense, so he sadly accepts the role of "harsh taskmaster," secure in the faith that he is thereby doing his duty toward his men and his country. This might be called the "John Wayne Dilemma." There is a grain of truth here about the undesirable consequences from the point of view of the Army of too close contact between officers and enlisted men. Given the American tradition of "equality," the possibility that officers could be seduced from the perspective of their superiors by primary group loyalties is one that cannot be discounted.

The discussion of primary group loyalties in *The American Soldier* is relevant in this connection. In their very complicated interpretation of combat motivation, the authors place great weight on this factor. Their basis for doing so is derived at least in part from incentives named by infantrymen themselves and summarized in the following statistics. Their replies to the query, "Generally, in your combat experience, what was most important to you in making you want to keep going and do as well as you could?" broke down into five main categories:

a. 39% emphasized ending the war
b. 14% emphasized solidarity with group
c. 9% emphasized thoughts of home and loved ones
d. 10% emphasized sense of duty and self respect

e. 28% emphasized miscellaneous factors, including patriotism, ideals, vindictiveness, etc.[8]

The assumption that this breakdown suggests the primacy of primary soldier groups could be challenged since *a* has really no group affiliative implications while *c* suggests the influence of another and perhaps closer small group—the soldier's own family. Only *b* with any certainty and *d* somewhat less certainty point to the role of the small informal group. Actually, the overwhelming evidence of the chart is that many soldiers in combat did not think about anyone or anything except their prospects for getting out of it—a feeling that is not easily dismissed.

Chapters Two and Three of Volume Two of *The American Soldier* present a detailed picture of the combat situation from which a few relevant points will be drawn.

In the previous discussion of the resentment against officers felt by enlisted men, no specification as to where these troops were located was included. It is worth noting now, however, that resentment against officers was least, and confidence in them highest, among front-line combat troops. The chart on page 366 of Volume One shows significant differences in attitudes toward officers varying with closeness to the front lines. This section suggests that maximum dislike of officers occurs among garrison troops and trainees while some kind of understanding between officers and enlisted men seems to grow out of the shared experience of combat.

These findings become more intelligible if we link them to the descriptions of combat units in Volume Two. Here we find that combat makes impossible the usual "caste" privileges of officers. Deference patterns become dangerous; salutes, special uniforms, and insignias are likely to bring concentrated enemy fire. At the front no one has consumer luxuries and available scarce goods are generally shared without regard to rank. Informal solidarity across caste lines is permitted

[8] Stouffer, *op.cit.*, Vol. II. Adapted from Table I, p. 109.

while the common purpose of getting out of the engagement alive overshadows most caste differences. Officers in combat have to lead from the front, a peculiarity arising from American conceptions of leadership and equality rather than from any inherent military necessities. Insofar as they do this, their risk is greater. Combat officers and enlisted men alike have a certain solidarity with one another which differentiates them from all soldiers who haven't been in combat. Mauldin recognizes this in his cartoon showing a bearded, unkempt combat captain and his equally unkempt enlisted driver in a jeep looking at signs on an Officer's Club restricting it to non-combat officers and at a Soldier's Club specifying that ties must be worn. The caption reads: "The hell with it, sir. Let's go back to the front." On the subject of relaxation of discipline, a cartoon showing a grizzled lieutenant playing cards in a front-line dugout with Willy and Joe, who look as beat up as ever, has the caption: "By th' way, what wuz them changes you wuz gonna make when you took over last month, sir?"

Then, one of the main features of these combat primary groups lies in their ability to incorporate low-ranking officers and thereby eliminate the more unpleasant features of caste. Yet this is one of the things that the caste system was supposed to prevent. There is a kind of paradox here which deserves closer examination.

If caste is strongest in training and garrison contexts and weakest in combat, there would seem to be some likelihood that its more important functions lie in the non-combat phase of Army life. Though there is no really intensive study of the training situation, scattered comments suggest that this may be the case. Chapter Two of Volume One, "The Old Army and the New," does report some responses to the training program during the early period:

"I wish the officers would treat us like intelligent adults. Men inducted into the Army are those who were independent in thought and action, in other words worked for their living. Maybe the old Army had men who signed up because of the easy life and lack of responsibility involved in the shaping of the future. I wish there would be less of the monotonous repetition. Men take less interest in their work, fool around, and consequently annoy the officers, making it hard on everybody. Treat a man like a nitwit and he'll finally act like one."[9]

"The men are usually kept in the dark as to what they are accomplishing, personally or in units, and questions as to the reasons for orders are barked down immediately."[10]

"Constant inspection of equipment prepared for this or that purpose simply to fill up time. In this connection the practice of encouraging soldiers to keep one set of equipment for inspection and another for use puts a wholly false emphasis and takes it off the necessity for having the equipment (clean)."[11]

"The Army is the biggest breaker of morale. The Army idea of class distinction between officers and men is all wrong. The Army does not take advantage of its man resources. Men do not like to be treated as if they were just toys and dogs for someone to play with. We are entitled to the respect we worked for and earned in civilian life."[12]

"The officers of my regiment take advantage of sheer rank by subjugating the private to their every whim and desire. Why should they (the privates) have to wait on their table, clean their dishes, roll and unroll their beds, and a million other things which every healthy man should be able to do for himself? We who were drafted for one year, and probably more, expected to devote that year in learning to soldier. Or was our conception of soldiering wrong?"[13]

These quotations and others in the same chapter show that Army training came as a shock to many of the draftees. Some of the things they objected to were matters of necessity. Rou-

[9] Stouffer, op.cit., Vol. i, p. 71.   [10] Ibid., p. 70.
[11] Ibid., p. 78.   [12] Ibid., p. 74.
[13] Ibid., pp. 74-75.

tine repetitiveness of training was explained by the Army as due to the fact that they all had to be similarly trained even though the draftees had different capacities and therefore teaching was pitched pretty close to the lowest common denominator. Again, some of the resentment in this early period arose because the draftees were better educated than the regular army non-coms giving them instructions and orders.

Taken as a whole, it is clear that Army basic training aroused deep insecurity and resentment in the soldiers exposed to it. In the first place, there is the fact that men in this position were stripped of their civilian ego-supports; property, clothing, family, and friends were discarded and a standard uniform with dog tags to match was provided impartially to all. The new world into which these trainees were plunged was an all-male world and therefore quite different from the civilian settings most of them left behind.

It is hard to reduce this complex emotional experience to simple terms. Perhaps the best psychological interpretation is that presented in a remarkable book, *Insight and Personality Adjustment* by Therese Benedek, dealing with the psychodynamics of all phases of military and veteran life. From a sociological standpoint, basic training served as a "degrouping" and "regrouping" agency so that the psychic unrest that it engendered was partly the result of the immense transitions that it did succeed in accomplishing. Americans from widely different class, religious, and ethnic backgrounds had to be stripped of their old identities and coerced to accept new military roles even though these violated some of their basic values. Since such ideological incentives as patriotism could not be depended upon, this training period had to establish new incentives. This is the point at which the motivational functions of the caste system enter. While ostensibly it appears to dampen the willingness to conform to military roles on the part of enlisted men who suffer from it, in reality it channels their motivation into one kind of military role—that of combat—while leaving those soldiers who do not serve in this fashion to confront the status agonies associated with caste inferiority.

In this context, the superabundance of caste "chicken" in training and garrison settings and its near disappearance in combat makes sociological sense. The trainees are exposed to vigorous attacks on their own self-esteem. Their civilian self-images in which they saw themselves as responsible adults are threatened by the Army's systematic program of disciplinary training in which officers treat them like recalcitrant children. Their most intimate personal habits, like shaving, hair-cut styles, body postures, and the like are subjected to intense scrutiny and regulation. Not only are their activities scheduled for the periods of time that they are on duty; even their leisure is subject to control. Thus, off-camp saluting was one of the main sore points between officers and enlisted men in many training centers. The caste system with its discipline for the enlisted men and privileges for the officers becomes the main focus for the resentment aroused by the Army's attack of the self-esteem of its trainees.

Since it is the officers who wield the power, acceptance by fellow trainees can hardly restore the shattered self-esteem of the enlisted men. In training units, with their limited life expectancies and complete domination by officers and non-coms, the emergence of powerful informal groups among the troops is most unlikely. At the same time, it is this period of training in which informal support is most desperately needed. Impulses for primary group satisfactions are actually exacerbated by the social situation. The troops are subjected to seemingly irrational disciplinary measures but are prevented from invoking primary group defenses against these measures by the social system in which they are bound.

With this threatening situation in mind, the considerable amount of resentment and anxiety expressed by trainees is to have been expected. The Army, on its part, leaves them very few channels for regaining their lost self-esteem. A few can themselves become officers and enjoy caste privileges at the expense of their less fortunate comrades. Others can begin the long climb through the ranks to get themselves whatever status and security can be gleaned as non-coms. Most, however, have to accept the necessity of serving in permanent units as privates and the only alternatives are whether they will go into garrison units or combat units. Though they have little control over this decision, there can be no doubt that most would prefer garrison duties to combat. But the effect of training is to make soldiers feel it necessary to say that combat units are more acceptable to them despite the risks to life and limb involved.

One reason for this is that trainees were exposed to the conception that combat units were places where their shattered self-esteem could be restored. There the officers ran the main risks while hated caste etiquette was dropped in favor of front-line comradeship. Combat men were looked up to by all other units. Their masculinity was confirmed beyond a doubt. This was especially the case in the training camp itself, where cadre who had seen combat were accorded special deference well beyond their nominal rank.

As the quotations from the Persian Gulf Command (a garrison unit) suggest, tension ran high in overseas garrison units over caste privileges. This situation involved an attack on the self-esteem of the enlisted men which they could modulate somewhat by considering themselves "lucky" not to be in combat. Furthermore, they were always able to gripe to each other and look forward to eventual return to civilian life, where caste inequities would be absent. Complaints about caste allowed garrison troops to discharge some of the guilt they must have felt at missing really hazardous duty, such as combat entailed, by making it seem as if they too were suffering.

There is really not enough data on the relations between training, garrison, and combat social systems to reconstruct the whole picture.[14] Hypotheses like those developed above remain suggestive rather than definitive. From the standpoint of a theory of communities, their main significance lies in the conception of the responses to military service forthcoming from Americans during World War II. Mauldin's cartoon showing Willie and Joe standing before a bulletin board with notices reading "Officers will eat at enlisted mess," "Saluting no longer required in Company area," "Regimental Dance in C.P. Building; Officers and Enlisted personnel invited" with the caption, "Looks like we're goin' into th' line, Willie," makes a good deal more sense if the reader knows that caste and combat didn't mix. It becomes even more intelligible if the foregoing theory about the ego-shattering consequences of caste discipline and prerogatives is taken into account. Without this background, another cartoon showing Willie carrying a lieutenant through a shower of enemy bullets on his back and captioned, "Don't mention it, lootenant. They mighta replaced ya wit' one of them salutin' demons" remains utterly incomprehensible.

Moving to a slightly more general level, it appears that the real meaning of the war for the men who fought it involved only slight identification with a national crusade or even with national defense. Nor were their attachments to small groups of buddies quite as important as the authors of *The American Soldier* occasionally suggest. The striking fact was the capacity of the American soldier to define the war, as he had the depression a few years earlier, as another "dip" or "temporary setback" in his personal and society's collective march toward a better future. His comparative immunity to other ideological reflection paralleled his immunity to concern

[14] A more comprehensive discussion of the social psychological effects of military training can be found in the article, "The Dissolution of Identity in Military Life," by Arthur Vidich and Maurice Stein, appearing in *Identity and Anxiety,* edited by Stein, Vidich, and White.

over the long-range goals of his corporate employers. This was the same condition that made small groups so important in industry; in a sense it was simply the lack of any larger groups whose purposes mattered. Even here, except under combat exigencies where survival itself depended on the closest cooperation in the combat team, our soldiers were far more interested in pursuing their private status goals through promotion than in retaining small group ties at the expense of advancement.

Hostility against the officers and caste prerogatives seem to have been the strongest feeling that Army life generated. But this too has its relation to earlier community experiences. Only a return to civilian life could redress the caste imbalance of Army life. The anxieties that coping with it entailed only added further incentive to the desire to get out of the Army so as to resume accumulating commodities again unhampered by caste constraints. The motives on which industrialism depends were thus accentuated by the artificial rigors of military service.

Other aspects of the hostility toward officers prove to be equally useful instruments of social control. The response to training camp "chicken," which challenges masculinity to the point of eliciting verbal expressions of preference for combat, was hardly contrary to the Army's wishes. Furthermore, the continuous griping about officers and caste deflected attention from more fundamental sources of dissatisfaction potentially arising from the lack of ideological conviction. It tied up the enlisted man with endless griping, which, if he had had any strong ideological convictions about the cause for which he was being asked to make major sacrifices, would have seemed less absorbing and consequential than it did.

Perhaps the only active response expressing resentment against the Army and military service was the minor "sabotage" committed by "goldbricks." Actually, these usually took the form of excessive time devoted to assigned tasks, a not

uncommon practice in large bureaucracies and hardly constituting open rejection of the Army or its authority system.
"Goldbricks" who didn't make more work for others were held
in high esteem by their fellow enlisted men, and exceptional
talents along this line usually brought considerable prestige.
Even this willful withholding of effort was not seen as necessarily entailing revolt except perhaps by overzealous officers.
It was really all part of a "day's work" and one way of guaranteeing that sacrifices would remain equal. Officers who enjoyed special authority and privileges therefore had to produce; enlisted men enjoyed neither and therefore they were
justified in "goldbricking" as much as they could. The same
attitude was taken toward the collective enterprise of the
Army as toward that of any industrial employer.

Like any other large-scale organization, the Army tried to
reduce its personnel to interchangeable "hands." Meaningless
discipline was imposed and enforced. The main response to
this depersonalization was a grim fatalism expressed in the
oft-repeated phrase, "Get your card punched" whenever a complaint was voiced. There is some humor or at least irony in
this response which implies that the Army which "issues" every
other kind of equipment imaginable also provides a special
"card" that you can have punched by the chaplain whenever
you want to register a complaint. It is the final bureaucratization of protest, only slightly less passive than the way in which
enlisted men in groups used to request that officers take some
remedial action when they had been left standing and presumably forgotten in inclement weather. Who can forget hear-
ing a group of adult males moaning softly as the rain poured
down on them, "Let's get the troops out of the rain"?

Military service then softened Americans for further bureaucratization at the same time as it left them with even less
sense of control over their destinies than had remained intact
after the depression. The status quest was not dimmed, only
further compressed into private life plans. Like the depression,

the war experiences had to be quickly blotted out before they distracted attention from the main task at hand. Our soldiers had dropped off the status ladder when the depression pulled it out from under them; they were not going to let themselves be dropped again. If they had any doubts now about the stability of the ladder as well as their confidence in their ability to keep climbing it successfully, these had to be ruthlessly suppressed for they might themselves interfere with the climb. Other doubts like those about the worth of the goals were so dangerous that they could not be entertained even in fantasy without risking the entire enterprise.

SUBURBIA: DREAM OR NIGHTMARE?

STUDENTS of the American community in the fifties have had their attention drawn to the suburbs much as the attention of their counterparts in the twenties was drawn to the slums. On the face of it, the reasons would appear to be quite different: the slum is the center of urban disorganization while the suburb would appear to be that of urban aspiration. Closer attention to the details of suburban life suggests, however, that it is actually the setting for the dominant "disorders" of our time. A popular and perhaps impressionistic study like Spectorsky's *The Exurbanites* starts with exactly this premise and provides a good deal of valuable information about the latest and most extreme form of suburbia.

This particular book is a good beginning point because it deals with an extreme instance of suburban development against which some of the more conventional forms can be measured with profit. Such communities as Park Forest, studied by Whyte, or Crestwood Heights, studied by Seeley, Sim, and Loosley, gain in perspective when placed against Spectorsky's work. The whole effort at developing a sociological interpretation of suburbia will rest upon the accumulation during the years to come of a series of studies of different kinds of suburbs. This chapter can move only a short distance toward such a goal. One purpose is to block out some aspects of suburban social structure and functioning which seem to hold promise as leads in the eventual development of a comprehensive typology of suburbs. Even more important, this chapter aims at viewing the suburb as a typical sub-community of the fifties, so that an examination of its problems and adaptative patterns will shed light on similar matters in other urban sub-communities which in their surface aspects would seem to be very different. As the slum was a main key to urban disorganization in the twenties, so the suburb may well be a key in the fifties.

The Exurbanites deals with a special category of suburbanites: the men who work in the communications industries in New York City. These men, in an effort to establish refuges for themselves from the urban "rat races" at which they are regularly employed, have infiltrated and finally taken over sections of several outlying counties including Westchester, Fairfield, Bucks and Rockland Counties. As luck (or perhaps their social situation) would have it, these people moved out to the "exurbs" in groups. Each exurban area tended to attract slightly different elements from the vast communications industry and Spectorsky has a very interesting comparative analysis of the Bucks, Fairfield, and Rockland settlements. His major analytic strategy is to link the social patterns of the sub-community to the particular occupational status and problems of its residents. He does this so consistently and so well that apparently chance differences become readily intelligible. If there is some question as to whether the communications industry in smaller cities would employ a sufficiently large and cohesive group to warrant the establishment of clear-cut exurban sub-communities, the metropolises are certain to develop some form of it. This is an instance where the special needs and problems of an occupational group predisposes them toward patterns of residential living that involve the establishment of spatial and social sub-communities.

Spectorsky pursues this theme from its initial definition wherein exurbanites are shown to be motivated by a reaction against their daily labors involving symbol creation and manipulation, through the effects of that peculiar social institution, the commuters' train, to their final entanglement with budgets that remain constantly beyond their incomes. Homes in the exurbs are designed to provide rustic comforts so that the "lucky" residents can balance their symbol-dominated work-days by thing-dominated leisure activities based on intimate contact with nature. At the same time, this contact with the wide-open spaces includes a sophisticated circle of neighbors and friends whose business and home lives are in the same state of precarious equilibrium.

Spectorsky describes the institutional system arising in each of the New York City exurbs in telling detail. Jobs in the communications industry are highly competitive, and apparently the criteria for judgment leave much to be desired as regards objectivity. The individual employee is therefore always subject to important uncertainties as to how his superiors view him. Since keeping his place in the exurban social system entails living well beyond his income at any given time and moving still further beyond it with each increase of income, he is always subject to terrible anxieties with no conceivable relaxation in sight. For that matter, as Spectorsky depicts the situation, the higher one gets, the more pressure one is subject to.

The job is not, however, the only source of pressure. In his chapter on women and children, the author points toward a number of family dilemmas arising from the isolation of the women in their rural strongholds. With all the expensive modern appliances that money can buy, life for these latter-day frontierswomen remains tied to schedules that are tremendously demanding. Their time is portioned out in taking care of their rambling houses, transporting their husbands to the trains and their children to school as well as to their play areas, and finally preparing themselves to greet their travel-and-work-tense husbands at the end of each busy day. Coming from careers and colleges themselves, these wives cannot help envying the professional and social outlets available to their husbands in the city. Boredom and impatience are the female's lot. She can focus on her children as an escape, perhaps experiment extra-maritally, or even take to drinking. Whichever way she turns, further frustration seems to lie in wait.

Perhaps the most painful chapter in the book is the one reporting on the leisure of the exurbanites. Liquor plays a pervasive role. Cliques abound, with the unfortunate result that the sociable account executive trying to escape from office problems finds himself thrown together with other account executives in the same situation so that the temptation to lapse back into office conversation and moods is inescapable. And since social life is coordinated with career goals, an unsuccessful party or excessive unpopularity can react back on job chances. Most exurban relaxation takes place at full speed: Sunday finds everyone tired from Saturday's round of vigorous family activities and evening partying. By this time the novelty of the children has worn off for father. Mother, trying to avoid spending this day as she does weekdays, arranges informal evening "dropping around." Spectorsky describes some of the party games that are occasionally played on these occasions. Their main purpose seems to be eliciting embarrassing expressions from the unwary while everyone else tries to release their hostilities without exposing themselves. The ingenious cruelty shown in this enterprise is amazing.

To interpret the strange suburban worlds described in *The Exurbanites*, the sociologist can begin by placing them in the context of a theory of community development. In part, these sub-communities are successors to Greenwich Village. Here the Pseudo-Bohemians who spent their early years in the Village while working their way up the hierarchies of the communications industry now try to recapture some of the values of their youth without sacrificing the comforts that accompany commercial success. This accounts for the heavy emphasis on "self-expression" in the exurbs. While most of them have given up anything more than a lingering dream of artistic success, they still seek to "express" themselves in their homes and life styles. Heavy emphasis is placed on original but good taste judged according to the most advanced ideas of the better fashion magazines.

By the time they have made the exurban grade, they are already too enmeshed in an over-extended budget to consider seriously returning to genuinely artistic employments with the consequent economic sacrifice that this usually requires. Yet Spectorsky notes in his last chapter the persistence of such fantasies, with the would-be "creative artists" cherishing the illusion that they will eventually write a great novel while their businessmen counterparts dream of going into business for themselves. As their obligations increase, the futility of these secret dreams becomes increasingly apparent. At the same time, they are confronted with the dilemmas of their more limited dream: escaping from the "rat race" to their exurban manorial life. Soon these expensive exurban homes, with the upkeep, improvement, and commuting problems they entail, begin to appear like a trap at least as crushing as the "rat races" for which they were supposed to compensate. Even the limited dream of exurban bliss is dissipated, while the secret dream of eventual creative accomplishment is even more traumatically shattered. Growing awareness of self-betrayal can be avoided by manic hyperactivity, by cynicism, or, for the introspectively inclined, by extended psychotherapy.

This condensed summary, harsh though it may sound, does not capture more than a bit of the detailed excoriation of the exurban plight which Spectorsky presents. The wish to escape from the harsher human consequences of bureaucratic role-playing has led a group of talented and economically privileged people to pattern a way of life that intensifies instead of alleviates the anxieties inherent in their occupational roles. Doctors in the community report an abundance of psychosomatic ills, including some like sexual impotence and frigidity which further disrupt already disturbed family lives. Perhaps because of the author's involvement and his hostility, the book fails to capture any sense of stable satisfactions that this round of life could convey. Some of the uncertainties and anxieties derive from the special early creative aspirations of communications workers; others seem to stem from the exceptionally competitive character of the industry itself. Both are reinforced by the fact that the skills sold are exactly those earlier intended as vehicles of creative expression. This is the modern version of the cycle described by Lynd, wherein desire for consumers' goods motivates people to play otherwise unrewarding production roles; the circumstances are a bit more complicated here, but the principle is the same. Middletown is not as far from Bucks County either in problems or social structure as the residents of both might assume.

A different kind of suburb, bearing only slight physical resemblance to exurbia, is the housing "development" or "project" which is making its appearance at the outer edges of most American cities. These new suburbs, most of them erected during and since the war, have standard ranch-type houses or, in lower-income brackets, standard box-type dwellings. Sometimes they are all built by the same firm, as in Levittown or Park Forest, the community studied by William H. Whyte. His analysis of this community appears as Part vii of his book *The Organization Man.* Since residence in these suburbs is usually only a temporary affair, Whyte has aptly called his study "The Transients." These are the stopover communities in which upwardly mobile junior executives house themselves while serving their apprenticeship. In Park Forest, about 35 per cent of the population turned over annually.

Whyte's study provides a penetrating glimpse into the patterns of social life arising among these transient neighbors. Park Forest is a planned community built in an outlying part of Chicago by a single entrepreneur. The houses are similar though not identical, since provision for standardized variation has been built into the plan. Since most of the residents are rather low in the occupational hierarchies, incomes are comparatively small and excessive expenditure frowned upon.

The goal is to hit just that dead level of spending which will demonstrate good taste without a hint of extravagance.

Park Forest differs from exurbia in that its residents are not employed by a single industry. They are distributed among the lower executive ranks of many corporate enterprises, including even the army. But this fact of diversity in kinds of enterprises does not mean that their work situations or problems are different in any important social respects. All must focus their attention on gaining approval from their superiors and all accomplish this by demonstrating their "flexibility and capacity to adjust." They are Organization Men and must display virtues of their tribe. Thus, the whole social life of the community becomes a training ground in "adjustment."

All of the institutions of the community arrange themselves around the needs of the corporation. Since business has learned the lesson that an efficient executive needs a happy home, family life becomes most important. The families of Park Forest learn how to look happy. Since the houses are close together and privacy all but impossible, this is not a part-time task but one that usurps most of the wife's energies and a good deal of the husband's when he is around the house. During the day the women have interminable "kaffee-klatches." Everyone is forced to participate, even the more retiring, and the groups themselves develop techniques for "reeducating" would-be laggards. This neighborliness, forced in part by propinquity, is encouraged by the feeling that the wish to be alone is somehow neurotic or childish. Fulfillment comes from group participation.

This slogan is imbibed by the children at their mother's kaffee-klatch and reinforced by their early training in the play group, nursery school, and finally grammar school. Getting along with others is the prime virtue and popularity its symbol. Much of the curriculum of the Park Forest school is devoted to "life adjustment" courses having this as their theme. Their high school exemplifies this best: over half of its seventy formal course offerings fall into this category. Parents support this trend, maintaining that the high school is primarily responsible for teaching their children to be good citizens and to get along with people.

Religion in Park Forest is also caught up in "togetherness." The Protestant denominations consolidated in a United Protestant Church and employ a minister who has learned to de-emphasize religious doctrine so as to avoid offending anyone, while at the same time providing plentiful opportunities for social activities which his transient parishioners need to give them a sense of stability. Administering the invariably extensive affairs and finances of this church provides these young executives with an opportunity to see the over-all operations of a large-scale organization and thereby does its part in training them for the responsibilities they will someday carry.

The social life of these Park Foresters embodies the same principle. Their parties must be modest but show some signs of originality in the kind of food served and entertainment provided. There are periodic waves of food fashions so that the housewife is never completely at loss for acceptable new dishes. Riesman's concept of "marginal differentiation" expresses well the kind of "individuality" that Park Forest encourages. Wandering too far from the margin brings penalties, as the housewife who was discovered reading Plato and listening to The Magic Flute learned. In this community, even the ways of "being different" are standardized.

With all of this "togetherness," the junior executive and his family must still look forward to the day when his promotion will both permit and require a rapid departure from Park Forest. Friendships cannot become so deep as to interfere with this shift upward nor can they become so deep that a friend's failure to move upward can disturb one's own upward rise. The "inconspicuous consumption" defined by Riesman is a form of "antagonistic cooperation" and this, after all, is the main operating mode of the upwardly mobile in large corporations. It is important to be "friends" with one's colleagues, superiors, and subordinates though the exact etiquette for expressing friendship with each differs considerably. Equally important, however, is the requirement that one not become so friendly with anyone that this interferes with effective competition. Friendliness is always subordinate to the goal of successful mobility.

So clearly do the transients appear to be the classical "other-directed" men of our time that it almost seems that Riesman must have developed his concept by examining their behavior. They can sink roots easily and tear them up with little difficulty. Anyone living next door can become as good a friend as anyone else. In Park Forest, spatial position determines friendship patterns to an incredibly predictable degree. Everyone is equal and therefore equally suitable for friendship. Religion, common interests, common backgrounds, or other likely factors are suspended in favor of proximity and adaptability.

This ready friendliness never reenters the lives of these transients with quite the same indiscriminateness as in Park Forest. Some of them apparently always yearn for such ready sociability even after moving to more expensive but more austere neighborhoods. They miss the continual "socializing" to which they have become addicted and feel uncomfortable in their new surroundings.

Crestwood Heights, a study of a Canadian upper-middle-class suburb by John Seeley, R. Alexander Sim, and Elizabeth Loosley, reports on the social life of the destination suburbs to which the more successful transients migrate. This is the first full-scale sociological study of a suburb. Brilliantly executed, it is a fit successor to the classic community studies.

In assessing the status of this research as a case study, the fact that Crestwood Heights is in Canada rather than America would seem to lessen its value for this theory. The authors do not dwell on the specifically Canadian characteristics of the community but rather view it as a typically North American phenomenon. In his introduction to the volume, David Riesman notes quite correctly that further specification of the influence of the national context might have proved worthwhile. When comparable American studies are available, a more careful socio-temporal location of these findings will become necessary. The limited reports on our suburbs now in hand suggest that the bulk of the important generalizations in *Crestwood Heights* do apply to similar American suburbs.

Crestwood Heights is a well-to-do suburb. Many of the residents are independent business and professional men, while the "employees" are usually senior executives at or close to policy levels. Houses are expensive and living even more so. One point of interest is the fact that this suburb is in transition from a predominantly or exclusively Gentile to a predominantly Jewish neighborhood. The processes of invasion and succession involved in this transition are never brought into focus, though they crop up at several points. Though this failure to deal with the community in terms of a process model is unfortunate, the inattention to Jewish-Gentile differences has an interesting rationale. The authors found that outside of church and a few ritual occasions, these differences were negligible with regard to social behavior. There is obviously some segregation within the sub-community. Jews and Gentiles both associate with members of their own faiths more than with each other and intermarriage, though it does occur, evokes considerable censure. The two religious groups may attend church on different days, worship a different god, and observe different rituals but the capacity of any of this to alter the deeper meanings of their lives is slight. Above all else, they are successful businessmen (or female appendages of successful businessmen) and their Judaism or Christianity remains conveniently in the background except on religious holidays or when intermarriage threatens. Characteristically, the main area in which concern over religion is expressed occurs in the high school, where fraternity or sorority "discrimination" does become an object of some controversy.

Seeley, Sim, and Loosley are more aware than most sociologists that their study could disturb the community when various members of it read the book, as they inevitably will. But while their preoccupation with the disturbance potential of this study is commendable, one suspects that the authors were more disturbed by their findings than the Crestwood Heighters are likely to be. In this purported "paradise" of mid-twentieth-century capitalism, they found a depth and degree of social disorganization that surprised and even frightened them. If the facts they report are true, one suspects that the success-driven men of Crestwood Heights will have little difficulty throwing off the effects of the study, though the women might react more strongly. Resistances to insight challenging fundamental values are notoriously powerful, so that the authors' concern occasionally seems a bit excessive.

This concern may, however, account for their own reluctance to draw together some of the sharper implications of their study. Thus their last chapter summarizes the implications for mental health education while only indirectly diagnosing the condition of the whole community. Actually they do point out the close integration between mental health programs and community institutions. A major point is the observation that mental health experts are constantly being called upon to give essentially moral guidance, but they never explicitly link this up with the breakdown of established norms in different areas of life. They call attention to the rise of a new priesthood of "mental health" and beautifully document its growing role, but somehow decline to observe that the supply and demand in this instance are both aspects of the same problem constellation. The main contribution of their book is the delineation of this constellation, their penetration beneath the surface façade of "health and happiness" which suburban families present to the world.

II. DEVELOPMENT

One suspects that much of the depth of the book derives from the authors' interest in mental health and their utilization of this interest as a lever for entry to the community. They worked out of the mental health agency in the school and in so doing were forced up against members of the community whose disturbed conditions had led them to consult the clinic. For the most part, these were families with "problem children" though adults in difficulty also must have come within their purview if only because disturbed children invariably have in the background disturbed parents. The parent-child relationship proved to be an important lead to the problems of this sub-community, one that naturally grew out of concern with mental health. And in this suburb it could hardly have been avoided.

The social structure of child-rearing deserves close scrutiny, since it is in itself one of the most problematic aspects of Crestwood Heights life. After a long description of the "confusions" that exist the authors summarize: "The whole area of child care and control is at present in a state of considerable flux in Crestwood Heights. Fathers may differ from mothers at almost every point in the complex socialization process. Families are divided between permissive and authoritarian ways of handling children, and so are individuals at different times or—worse—the same time."[1] At the same time, there is a tremendous desire that the child be "normal" for his age so that signs of difference send the family off to the experts. Even the normal socialization routines have been turned over by confused parents to nursery schools and other secondary agencies. The child is expected to learn independence at an increasingly early age:

". . . The busy mother, who must run the house without the aid of a maid, entertain for husbands and friends, attend

[1] Seeley, John R., Sim, R. Alexander, and Loosley, Elizabeth W., *Crestwood Heights*, Basic Books, Inc., New York, 1956, p. 203.

[210]

meetings, and have 'outside interests,' is literally compelled to ration strictly her physical contacts with her child. Thus in early infancy preparation continues for achievement-in-isolation, for the individual pursuit of materialistic goals in which human relationships must often be subordinated.

". . . There are no clear-cut norms as to what is expected, no absolute measuring sticks to serve as a guide. The child who is two this year is not expected to be exactly like the child who was two last year; attitudes are required to be always flexible and 'expectations' are constantly changing. What is constant is the expectation that expectations will change.

"Because of the mother's uncertainty, her sense of responsibility towards the child increases. Because practices constantly change, she tends to seek advice and help from sources outside the family. A pediatrician will advise in matters of health. A vast amount of printed material is available to the highly literate Crestwood mother. In the women's magazines she is confronted with innumerable articles on child care written by convincing experts. These present opinions on topics ranging from the proper nutrition for her baby to the best means of 'developing in a positive way the mental health of her child.' Many of the articles are by professionals, but she is also bombarded by the statements of other persons, representing commercial and business interests: the toymakers, the manufacturers of special foods and clothing for children. Each has a word of advice for the mother—or of censure. Each suggests something she should 'buy,' something which is calculated, if not guaranteed, to contribute to her child's welfare. There are also her friends, with whom she can exchange notes on the height, weight, walking ability and so on of her children and with whom she can search for solutions to the 'problems' which confront her.

"These innumerable recommendations do not, however, supply the mother with definite instructions. Their often contradictory directives only add to her uncertainty amid the changing patterns of her culture."[2]

Thus, the suburban mother is faced with a plethora of different and often conflicting norms exactly at a time when she feels a heightened sense of responsibility but has little objective opportunity to raise her child. Her other roles demand time and can hardly be abandoned because her husband's success depends on them; hence she has to be willing to turn the child over to secondary agencies. Her willingness to do this is enhanced by the breakdown of child-rearing norms since the people she is handing her child over to are the ones presumed to be competent at determining these properly (i.e., scientifically).

Beneath all Crestwood social patterns we find the same harnessing of life to occupational success explicitly adopted in Park Forest. Male careers do not involve quite the desperate struggle that Exurbia requires, but they do remain the anchoring point for the family's activities as well as the single factor on which continued residence in this privileged suburb depends. So the career of the male is focal, with female careers being subordinated to motherhood. This is usually painful for the serious woman, leaving her with even more ambivalence toward the already fragmented maternal role. Male career cycles have a terrible inevitability with performance peaks usually being reached in the late thirties only to taper off in the fifties and with virtual obsolescence arriving at an ever-earlier age. In addition, the limitations of the "peak" have to be absorbed, since no ceiling on ambition is ever quite acceptable in this sub-culture. Children seeing the career struggles of their parents soon become concerned with their own futures, yet parents cannot always arrange to pass on the fruits of their economic success to their children. All too often these fruits are consumed so that the child who has led a sheltered

[2] *Ibid.*, pp. 88-89.

existence with all of his material wants satisfied must still carve out his own competitive career when he leaves home.

In a brilliant chapter the authors analyze the suburban home to show how it functions as property, as a stage, as a focus for family living, and as a nursery. It is a tremendously expensive and lavish enterprise. The Crestwood child soon learns that the display functions take precedence over his impulses whenever the two conflict. The goal is "artless" staging of whatever setting the woman of the house has decided upon as a basis for justifying her claim to having "good taste." She makes her choice with the help of fashion magazines and interior decorators. Like child-rearing patterns, styles in home furnishing change periodically, giving the housewife another fashion to keep abreast of if she wishes to hold on to her position in the community. With all of its lavishness, the house never really captures the deeper affections of its members, for they are all aware that they would have to leave in case of economic success or failure. But even more important, they know that they will in due time all leave—the children for homes of their own and the parents to smaller houses in which to spend their "old age."

One of the lessons soon learned by the child is the appreciation of properly displayed property. He begins to feel this as a source of ego-enhancement so that new items are always preferred to old. As the child accumulates property for his own room, the advertising theorem that happiness comes from acquisition is painlessly inculcated. He watches his mother light up when she gets a shiny new kitchen appliance, so that cooking for her, as the advertisers would have it, does take on new meaning. But as the appliance ages and loses its capacity to bring appreciative responses from the neighbors it has to be quickly replaced by another. Unfortunately, the accumulation of appliances can never render cooking permanently meaningful as long as the woman is unsure of its relation to her feminine identity and of the value of

[213]

that identity itself as well. Like his mother, the child is forced to substitute property acquisition for deeper role satisfactions.

Learning to distinguish display functions from utility functions in the house is the beginning of the formation of a marketing self-image in which the child sees himself as a "commodity" to be displayed and sold. The child soon learns how strongly his parents feel about his capacity to keep up with his peers and maintain a presentable front. Like it or not, he must dedicate himself to this enterprise so that the appearance of success is as important as the reality. Eventually the distinction between the two breaks down. Children manipulate their personalities according to the latest theories which they pick up almost as quickly as their parents. They know the importance of appearing "well balanced" in order to sell themselves to the primal customers, their parents, as well as to their peers and teachers. This learning process is abetted by the media of communication, so omnipresent in the modern home, which tell them what kind of packaging they require to keep in style.

Life in the home is chopped up into time intervals all carefully scheduled to regulate the diverse affairs of the busy family. Very often the schedules of individual members are such that the whole family almost never spends time together during the week. Compensatory efforts are made at cramming intense "family life" into the equally busy week-end leisure schedules. The rhythm of the office has penetrated the home so that even television programming has to be taken into account: ". . . the resultant schedules are so demanding that the parents feel themselves constantly impelled to inculcate the virtues of punctuality and regularity in themselves and the child, at meal hour, departures for picnics, and such occasions. Being on time for school becomes more important than eating breakfast. The intimacy of primary relations can be punctured easily: the flow of after-dinner conversations will often be broken clearly and sharply at nine by a vigilant host who has scheduled a television viewing as part of the hospitality of the evening; a chattering group of students will be divided neatly into two halves by an ongoing bus (four minutes late), half boarding the vehicle, the remainder continuing on foot; family dinner is interrupted by long-distance phone calls for father. The secondary institutions thus emphatically affect the rhythms and patterns of family life, and the family with its generalized function is hardly in a position to resist the outside institution with its specific function, which, within limits, permits it to demand the individual's participation. To take a perhaps extreme instance, if a television program for children changes its schedule without warning, the family meal hour may have to be changed, unless the adult members are prepared to accept enforced silence and semi-darkness."[3] In school, students learn to parcel out their time in the same fashion as their fathers and mothers. There is little room for spontaneous association or activity in this "time conscious" community.

Viewed more comprehensively, life time is transformed into a disposable commodity. As in Muncie, there is a focus on the future but the certainty that this future will be better is disappearing and the emphasis on youth also noted by Lynd seems to have become more pervasive. The realities of biological maturation with its accompanying crises are almost ignored in favor of celebrations of competitive achievement: "The peculiarities of perspective, tempo, and rhythm in Crestwood Heights have interesting results in the growing-up pattern through which the child passes. . . . In many of our experiences in Crestwood Heights, expression has been given to fear and uncertainty respecting the future. This has taken various forms but the constant theme is the number of things to be done and the shortness of time remaining. For them, time appears to be 'running out.' . . . In a group of eleven- and twelve-year-old students, discussion really waxed hot over kiss-

---

[3] *Ibid.*, p. 65.

ing games. The girls say they are really not in favor of them, but Guy said there was nothing bad about them and objected to the discussion. Someone countered that they were too young for such games. Then someone wanted to know if they should wait until they were thirty-five before having kissing games. Jerry argued that *now* was the time for them, since when you got older you didn't play them, and if 'you don't do it now, when will you!' There is nothing unusual about the game, but the justification offered (that this is probably the last chance) does seem unusual. Indeed, if this trend were fully generalized it would entail the collapse toward youth of the entire life experience. Perhaps this is what happens, for if children are rushed to emancipation and adulthood there is little left for later years. Parents in Crestwood Heights frequently complain of this trend, but state that they are powerless to combat it since other children and other parents set the pace. This feeling is reciprocated by other parents, so that we have an example of an inflationary market system where normal controls are inoperative."[4]

Other evidence for this condition follows:

"Mrs. A has taken her four-year-old daughter to the hairdresser for a permanent . . .

". . . The Nursery School is taking two-year-olds now . . .

". . . Ten-year-old children in our school have dates, go to movies in twosomes, and have evening dresses.

". . . Our twelve-year-old boy wants to buy a tuxedo. He says everyone else in his class is buying or renting one for the Prom."[5]

All of this and more constitute the pattern of "rushing at experience." The result is a mockery of adult patterns of independence, competition, and sex because they are adopted long before any of them can possibly be integrated into the growing personality: "This rushing at experience has consequences for the child as he gradually forms a conception of himself as an

[4] *Ibid.*, pp. 69-70.   [5] *Ibid.*, p. 70.

'aging individual.' The boy who rides a horse, wears long pants, and has a girl friend before puberty, has lost some of the means of validating his biological manhood at the moment in his physical and social maturation when such signs are most needed. The boy's changing voice and the girl's first menstrual period are not accorded the public attention that one might expect from a culture where sexual matters are so often exposed to discussion. In contrast, there is, of course, no lack of ceremony in the areas where individual achievement and competition within the group are encouraged. These activities, which are often of a vigorous nature, and the rewards which attend them, together with opportunities for controlled eroticism (such as dances and mass behavior at football games), help to prolong adolescence and make it not simply tolerable, but 'exciting.' "[6]

Little can be added to this fine interpretation except to underscore the fact that the high estimation of youthfulness is accompanied by a tendency to transform it into a replica of adulthood. A more complex conception of the relation between adulthood and childhood which recognizes distinctive differences as well as continuities has given way to a confusion between the two. This confusion ramifies through the life cycle so that every age level is forced to feel ambivalent about the stage it is passing through:

". . . Up to a certain age, apparently, the passage of time is thought much too slow. Birthdays seem far apart, an impression which is fed by the other ceremonials of aging which occur during the year. After the crest of the golden ridge has been reached, the passage of time is viewed with regret; and birthdays, which are observed now with reluctance, seem too closely spaced. To those on the young side of this ridge, which is ascended so slowly, it is a compliment to say, 'You are big for your age'; 'You look more than sixteen.' Once on the other side, the descent is all too swift; the compliment is rather,

[6] *Ibid.*, pp. 70-71.

'You can't be that old'; 'You don't look a day over twenty-four.' The age of twenty-four among women at least seemed to be the 'golden age'; for the male, the golden age is somewhat later, perhaps thirty-three. Up to that point, male youth seeks to be disguised behind a moustache or horn-rimmed glasses. Soon after, a man may find it difficult to alter his career line, or find a new practice or firm to work with, unless the move is clearly within the definition of career advancement."[7]

Naturally these images of male and female "golden ages" are reinforced by the advertising and other mass media messages, most of which refuse to recognize the worthwhile existence of anyone more than a few years off in either direction:

"For both sexes, despite these efforts to speed the clock, or, after a certain age, 'to make time stand still,' despite diets, cosmetics, and facial surgery, age creeps on. A critical period for the male, as it certainly is for the female, is the 'change of life.' Surrounded by more secrecy than is puberty, except when it is the subject of not too gentle joking, this period threatens, sobers, and frightens.

"The crisis surrounding menopause is, however, more serious for the female, while the next general crisis, which comes at retirement, is almost exclusively a problem for the male."[8]

It is important to notice what is happening here. The human life cycle has been leveled by a cultural system which ignores the values that can be found in various ages. By refusing to recognize the major life transitions in such a fashion as to render both prior and later stages meaningful and worthwhile, Crestwood Heights condemns its children and its adults to permanent anxiety in regard to their present identities as well as those of the future. And even if this anxiety can be channeled into competitive achievement, the price would seem to be permanent emotional immaturity.

There is more discussion in *Crestwood Heights* about the dilemmas of suburban socialization than on any other theme.

[7] *Ibid.*, p. 71. [8] *Ibid.*, pp. 71-72.

One of the main sources of concern to people in the community, this problem provides the sociological observers with profound insight into the disorganization that exists. It is clear that the routinization of the home is related to the suburban round of life which thrusts an overwhelmingly heavy schedule of activities on all the members. Male preoccupation with business success is matched by female preoccupation with house-wife-community success, while the child turns to his school and peer-groups with a similar perspective. These preoccupations can easily become compulsive outlets for evading anxieties generated by a pervasive failure at performing elemental human tasks. While uncertainties about parenthood occupy the center of the stage, there is enormous evidence that this reflects deep uncertainties about *all* the roles through which the human drama must be played out.

Schools are the main institutional mechanism in Crestwood Heights for coping with socialization. The child is handed over at an increasingly early age. Directives from school-affiliated experts exert their influence even earlier. The centrality of the schools in Crestwood Heights is exemplified by their size and spatial prominence. The authors contend that the community actually grew up around the school and suggest that its status can be compared with that of the church in less secular societies. They present a survey of the growth of school-centeredness in our society and follow this with a penetrating discussion of schools and education in Crestwood Heights. From the standpoint of community theory, the rise of progressive education and mental health has to be seen as a product of exactly the social processes that we have been analyzing:

"A highly interdependent and complex urban society, however, demands a more flexible, 'co-operative,' and at the same time, individuated human being to function in it. Respect for individual personality, the necessity so to use the child as to assure a constant flow of suitable personnel for industry and business, and insight into the role of early experience in forming the personality structure of the child—all have played a part in directing the concern of the school towards the mental health of its pupils. The older idea that the efficacy of education came, to some degree, from its difficulty and unpleasantness, gave ground to the idea that 'experience,' gained in free exploration of the environment, is the true educative force. A great paradox at this point confronts the school. The child must be free in accordance with democratic ideology; but he must, by no means, become free to the point of renouncing either the material success goals or the engineered cooperation integral to the adequate functioning of an industrial civilization."[9]

This "paradox" unfortunately can only be resolved in one way, a façade emphasizing growth and freedom with the actuality of heavy pressure toward competitive functioning always dominating the background. This coincides with the family situation where most of the things supposedly done *for* children actually are done *to* them. Lacking any opportunity to develop a sense of genuine selfhood, they are forced alongside their parents into the competitive scramble:

"To succeed in modern, urban society, the child must learn to maintain both competition and cooperation in a delicate balance of forces, and he must develop this balance through the learning situation itself. More exactly, he must learn a kind of covert competition, much more strenuous to keep up than open competition or abandonment of competition altogether: he must compete but he must not *seem* competitive. The school deals with the dilemma by overly 'promoting' cooperation (for example, by adjusting the teaching program and methods so that group experiences replace 'individualistic' learning) and by covertly 'tolerating' competition (for example, by retaining the system of competitive examinations and marks, modified to some extent through the years)."[10]

[9] *Ibid.*, p. 229.   [10] *Ibid.*, pp. 229-230.

Reflecting in part the uncertainties of the parents, the school now finds itself in a position to claim to resolve these uncertainties. It offers educational ideologies that promise to develop the "characters" of its charges, and programs that implement the promise though not necessarily in the direction desired. There is a proliferation of "human relations" teachers and courses so that orthodox subject matter, as in Park Forest, takes a back seat. The teachers are also caught in a trap since the official ideology prescribes "democratic" relations with their superiors and students, while the realities of school functioning belie this. Student confusion about authority is described:

"The pupil, in his turn, finds his relationship to his teacher equally subtle and elusive. On the one hand, he is expected to submit to the teacher's authority, whether it be traditionally autocratic or expressed in the current permissive terminology and technique. Whatever the teacher's concept of discipline, the pupil knows that this is the adult in whose charge he will remain for a considerable portion of the school year. This teacher—not he, nor his fellow pupils—is the one who will decide whether or not he 'passes.' Even his parents have not this most particular power to decide his fate. At the same time, and increasingly as he rises in the school, the pupil is encouraged to combine with this authoritarian adult image, the image of the teacher as a friend. It is as hard for the pupil to blend these images convincingly as it is for the teacher to believe that everyone is his equal in the hierarchical structure of the school. It is, moreover, hard for the school to recognize clearly the difficulties inherent in these conflicting definitions and to accept the concrete limitations imposed by the teaching situation. The strain engendered by proceeding as if these tensions did not exist is particularly evident in the classroom and markedly so in the higher grades. Yet the children who adjust well to this climate, as the majority appear to do, probably acquire by that very adjustment the training essential to enable them as adults to maintain quasi-intimate relations with people to impersonal situations—such as, for instance, the business man who feels he must treat all his customers as 'friends' (and perhaps also his 'friends' as customers)."[11]

The relationship to authority at school duplicates the confusions of the home, embodying the two dominant but contradictory directives toward "friendliness" and toward manipulation of others in the quest for success. The degradation of the concept of friendship imposed by this kind of educational system unfortunately fits all too well into the suburban-bureaucratic round of life. The social clubs to which Crestwood adults and children belong also reflect the enduring competitive preoccupation with status, though here the coating of "maturity" thins out even more than in the schools.

Crestwood schools are administered and staffed by a crew of college-trained educators and "human relations" specialists whose influence grew as family confusion mounted. Life in a complex scientifically inclined suburb like this one required the services of a great many experts on all kinds of matters: ". . . The once relatively simple matter of feeding the family requires the services of domestic science specialists; consumers' guides; hygienists and health teachers in general; pediatricians; specialists who can advise on the meaning of food and food preparation in the maintenance of happy married relationships; and communications experts, journalistic and novelistic, who will give clues as to what 'the best people' are *really* eating, regardless of all the foregoing considerations . . ."[12] Much of this desperately sought expertise requires some training in the social sciences. We observe efforts at applying social science knowledge that is only loosely confirmed in response to urgent demands arising from the confusions of suburban life. Some institutional mechanism had to emerge to meet this need. Applied social science then has become a vital part of suburban

[11] *Ibid.*, p. 271.   [12] *Ibid.*, p. 362.

functioning since it is the branch of science which claims to have knowledge about the vital life problems at the "collective, social or interpersonal" levels, on the one hand, and on the more intimate "intrapersonal or individual" level, on the other. Marriage counselors, psychotherapists, aptitude testers, child guidance experts, public relations men, and other related occupations are familiar fixtures in the suburban milieu.

Crestwood Heights presents a brilliant picture of the social system in which these experts and their clients are enmeshed. Starting from anxieties deeply rooted in the socio-cultural situations of their clients, social science "experts" are called upon to give assistance with symptomatic problems while leaving untouched the larger community context in which these problems arise. Sometimes the "expert" has access to a highly systematized body of techniques for providing help, as does the well-trained psychotherapist; all too often, however, he has to cope with problems going far beyond the limited practical or even theoretical development of his discipline. Aside from technical uncertainties, the expert's position in the bureaucratic structure is often unclear, which, of course, adds to his insecurities. On the other hand, in order to evoke favorable responses from his clients and superiors, he is forced to act as if his understanding of and control over the situation is much greater than is actually the case. This opens a wedge for manipulation of expertise along magical lines, a procedure that is itself reinforced by the uncertainties. A series of failures will eventually drive clients toward new sources of expertise. Since alternative theories are available in social science, new experts and programs can readily be found so that a regular succession pattern among experts appears. As any given expert expands his clientele, he is forced to dilute his theories by dropping complexities and qualifications. This brings the point of eventual disillusionment closer by making disparities between actualities and claims less susceptible to magical concealment. Furthermore, mass communication messages can

[223]

short-circuit the expert-client relationship by delivering contradictory messages as well as by heralding new theories to undercut current local vogues. Seeley, Sim, and Loosley note the emergence of "specialists" attached to functions rather than theories, who manage to maintain their positions by sharply limiting their scope, and carefully distinguishing themselves from the transient "experts." These men usually entrench themselves in bureaucratic structures like clinics and schools so as to ensure their place in the community.

The important influence that the "human relations" experts exert on this sub-community derives from their ability to shape the beliefs of the women. The men pay less attention even though there are signs that they too are beginning to listen—e.g., the utilization of "human relations" theories in both the production and marketing phases of many businesses. The inclination toward "human relations" theory shown by the women is rooted in their tendency to start from value themes emphasizing individualism, determinism, perfectibility, and emotionalism as well as by their greater exposure to the life problems in regard to which these experts are presumed to have competence. The authors summarize differences in belief systems between men and women as follows:

"The men seem primarily concerned about the preservation of life against destruction, and they feel and believe accordingly. The women seem concerned about the creative and elaborative processes, and they believe and feel accordingly. The men attend to the *necessary* conditions for living; the women to the conditions that would make life *sufficing*. The men are oriented to the biological and social substratum, to minima; the women to the social and psychological superstratum, to maxima. The men are concerned with the prevention of positive 'evils'; the women with the procurement of positive 'goods.' The men live psychologically in an emotional climate of scarcity requiring the close and calculated adaptation of means to ends; the women, correspondingly, live in a climate of abundance requiring the wise selection and utilization of the riches available. The men are for prevision—and provision accordingly; the women for vision—and enjoyment as of now. The men are sensitized to necessity; the women to choice. Compulsion, the vis a tergo, the drive from the past press with more weight on the men and order their behavior; yearning, 'final cause,' the pull of the future, lure or govern the women. Rousseau speaks more clearly for the women; Hobbes for the men."[13]

This split between the men and women is paralleled by the split emphasis in the school between "self-realization" and "grades" and leads to the inference that the child is confronted with the same polarity in his home. Since the behavioral emphases in the home as in the school fall on the side of the "success" values, there can be no real choice as to the final resolutions; but by differentiating the values along sex lines, the already formidable difficulties with respect to formation of adequate sexual identities are further compounded.

As a classical community study, *Crestwood Heights* contains many important points. For purposes of fitting it into this theory, a few comments on its relations to other suburban studies might be in order. There is evidence from Park Forest that the ideological differences between men and women reported in Crestwood Heights are in the process of being reconciled in suburbs where the sway of bureaucratic careers is more firmly established. Thus, transient husbands would be at least as likely to express collective ideologies as their wives. After all, these ideologies are becoming dominant in many large corporations. "Cooperation is the Keynote." The Crestwood males, being more independent, do stress traditional competitive values while their junior executive counterparts have picked up more "psychologically sensitive" motivational grammars.

Exurbia, Park Forest, and Crestwood Heights are only three

13 *Ibid.*, p. 393.

versions of a sub-community species which clearly contain other variants. They share the dedication to status and status-fronts along standardized "individual" lines; although the exurbanites would underscore "individual," while Park Forest emphasizes "standard" criteria more heavily. Perhaps the transients will wish to differentiate themselves more than marginally once they have established themselves in a more permanent suburb, but the question remains as to whether they will still be capable of doing so. The "individuality" of the exurbanites is purchased at a tremendous price—subjecting themselves to jobs that confine all opportunities for genuine expression within the bounds dictated by commercial necessities. Their "individuality" all too easily turns into bitter eccentricity in the face of these commercial pressures.

In all of these suburbs, children are at the focus of attention. Neither the exurbanites nor the Park Foresters seem to confront as many difficulties in deciding how to rear their offspring as do the Crestwood families, but all three share a willingness to abdicate in favor of the schools. Religion seems trivial in all three, turning into one more status symbol counting well below most others in the all-important secular realm. The growing influence of psychoanalytic doctrines and experts characterizes all three, although the exurbanites undoubtedly form the vanguard in this respect. Further interpretation of suburban social systems is found in Chapter 12. It seems appropriate to end this section, devoted to examining the effects of urbanization, industrialization and bureaucratization in American community life, with a chapter on this spreading sub-community.

PART III. PERSPECTIVES

INTRODUCTION TO PART III

THE purpose of Part III is to gain perspective on the contemporary community by studying it in light of selected anthropological and psychoanalytic findings. These disciplines draw on observations and interpretations of human life in the broadest possible terms. They throw the distinctive ordering principles of modern community life into high relief and provide conceptual tools for assessing the psychic and social costs that American communities exact in return for the opportunities they offer. Psychoanalysis and anthropology deepen our approach to community organization by detecting qualities and dimensions of human existence that sociological theories tend to neglect.

Chapter 10 draws on the work of several anthropologists, including Paul Radin, Robert Redfield, Ruth Benedict, and others, to develop an image of primitive community life which stands in sharp contrast to modern patterns. Primitive emphasis on ritual transitions throughout the life cycle is contrasted with the fragmentation of the life cycle in suburbia. The effects of this fragmentation on symbolic functioning and individuation are suggested.

Chapter 11 draws on psychoanalytic theories—especially the work of Sigmund Freud, Erik Erikson, and Harry Stack Sullivan—to explore the psychic ramifications of these discontinuities in cultural conditioning. Freud provides a framework for studying the effects of developmental "fixations" on adult functioning. Sullivan and Erikson interpret some of the culturally normal "fixations" which pervade modern life.

Chapter 12 draws together the threads of the entire volume by applying the psychoanalytic and anthropological perspectives to the interpretation of community processes in suburbia. The historical development of American community patterns is traced and analyzed. A recent study of a small town confirms many of the interpretations advanced earlier and sharpens our conception of the split between ideologies and realities in modern community life.

ANTHROPOLOGICAL PERSPECTIVES ON THE MODERN COMMUNITY

SOCIOLOGICAL theories about community development usually contain historical assumptions as to the character of earlier communities. If the theory aims at encompassing the broadest sweep of human history, the theorist has to include anthropological materials. In the best tradition of social-political theorizing, starting perhaps with Hobbes and Rousseau and moving through Marx, Spencer, Comte, Maine, Toennies, Veblen, Durkheim, and Weber, there is a serious effort to come to terms with available ethnographic materials. The classical theorists have been eminently responsible in this connection; for example, Maine's status-based social order, Toennies' Gemeinschaft, or Durkheim's mechanical solidarity all offer conceptions of primitive social structure which constitute important contributions to the development of both anthropology and sociology.

In this connection, Robert Park's sociological theory of urbanization contains many important leads for anthropologists. While his own work focused on the modern city, against the background of the rural community, his analysis of city life recognized the importance of folk cultural elements as a basis for important urban sub-communities and as a source of cultural vitality. Park's essays in *Race and Culture*, written throughout his long, active career, reflect an enduring concern with the effects of acculturation on the folk culture of the American Negroes. Robert Redfield recently employed Park's distinction between the technical and the moral order in community life as the central structural concept in his synthetic work *The Primitive World and Its Transformations*. Redfield's later volumes, *The Little Community* and *Peasant Society and Culture*, advance our understanding of this distinction considerably and point the way toward developing concepts for analyzing intermediate types on the folk-urban continuum.

There are signs that modern anthropologists are beginning to find more to admire in the folk societies they study than has been fashionable over the past few decades. The recent series of BBC lectures published as *The Institutions of Primitive Society* containing contributions from several leading British social anthropologists reflects this trend. It is by no means a form of primitivism or eulogy of any "noble savages" but rather proceeds with full consciousness of the special virtues of modern civilization. It seems to be a complex reaction on the one hand to the urgency of the immediate difficulties of that civilization, coupled with awareness that these difficulties can be illuminated when studied against the background of primitive societies where they do not appear. Anthropologists now seem increasingly willing to let their research throw light on the crises of our own civilization. They no longer feel that this necessarily entails abandoning scientific objectivity or choosing between primitive and modern ways of life.

Edward Sapir in a classic essay, *Culture, Genuine and Spurious*, first published in the early twenties used his familiarity with primitive society as a baseline for commenting on contemporary culture. While there is much that can and has been justly criticized in his effort, it certainly does block out some of the main considerations:

". . . a genuine culture refuses to consider the individual as a mere cog, as an entity whose sole raison d'etre lies in his subservience to a collective purpose that he is not conscious of or that has only a remote relevancy to his interests and strivings. The major activities of the individual must directly satisfy his own creative and emotional impulses, must always be something more than means to an end. The great cultural fallacy of industrialism, as developed up to the present time, is that in harnessing machines to our uses it has not known how to avoid the harnessing of the majority of mankind to its machines. The telephone girl who lends her capacities, during the greater part of the living day, to the manipulation of a technical routine that has an eventually high efficiency value but that answers to no spiritual needs of her own is an appalling sacrifice to civilization. As a solution of the problem of culture she is a failure—the more dismal the greater her natural endowment. As with the telephone girl, so, it is to be feared, with the great majority of us, slave-stokers to fires that burn for demons we would destroy, were it not that they appear in the guise of our benefactors. The American Indian who solves the economic problem with salmon-spear and rabbit-snare operates on a relatively low level of civilization, but he represents an incomparably higher solution than our telephone girl of the questions that culture has to ask of economics. There is here no question of the immediate utility, of the effective directness, of economic effort, nor of any sentimentalizing regrets as to the passing of the 'natural man.' The Indian's salmon-spearing is a culturally higher type of activity than that of the telephone girl or mill hand simply because there is normally no sense of spiritual frustration during its prosecution, no feeling of subservience to tyrannous yet largely inchoate demands, because it works in naturally with all the rest of the Indian's activities instead of standing out as a desert patch of merely economic effort in the whole of life."[1]

This quotation does not sum up Sapir's whole argument though it reflects the strengths and weaknesses of his position. His aspiration toward individual fulfillment *through* rather than despite community roles stands as a philosophic contribution of the very civilization which denies its realization. Con-

[1] Sapir, Edward, *Selected Writings of Edward Sapir*, edited by David G. Mandelbaum, University of California Press, Berkeley, 1949, pp. 315-316.

temporary existentialist critiques like those of Karl Jaspers and Gabriel Marcel elaborate these same points with similar emphasis on the fragmentation and despiritualization of modern life. Unfortunately, their grasp of the phenomenology of this experience remains as impressionistic as the "sense of spiritual frustration" and "subservience" that Sapir imputes to his telephone girl. Yet all three point to an important problem for the community sociologist requiring historical and anthropological perspective to conceptualize properly.

Running through this essay as well as Sapir's later articles on the interplay between culture and personality is his profound emphasis on the requirement that any society should provide its members with meaningful life activities through which they can grow and express their full individuality. It is this emphasis that distinguishes him from most recent "objective" students of the relation between culture and personality, and which gives his writing on this subject much greater depth than this later work. Ruth Benedict, ardent exponent of cultural relativism though she may have been, frequently adopted a similar perspective. Her familiar article, *Continuities and Discontinuities in Cultural Conditioning*, which is probably as widely known among sociologists as any other single piece of anthropological writing, is an important case in point. Here she shows that our society forces sharply contrasting roles over the life-cycle progression from child to adult as regards responsibility, authority, and sexuality without providing any significant transition rituals to bridge these discontinuities.

Her main thesis is in itself rather obvious, but it gains its special force from the brilliantly chosen illustrations of contrasting instances among non-literate cultures where these role disparities are either rendered continuous throughout the life cycle or bridged by elaborate rituals. It is hard to forget her description of the Papago grandfather's respectful patience while his three-year-old granddaughter closes a heavy door or

[233]

the reciprocity expressed in primitive kinship terminology where the same term applies to father and son as our term cousin applies to equal participants in a paired relationship. Her comments on comparative sexual training deserve mention: "If the cultural emphasis is upon sexual pleasure the child who is continuously conditioned will be encouraged to experiment freely and pleasurably, as among the Marquesans; if emphasis is upon reproduction, as among the Zuni of New Mexico, childish sex proclivities will not be exploited, for the only important use to which sex is thought to serve in his culture is not yet possible to him. The important contrast with our child training is that although a Zuni child is impressed with the wickedness of premature sex experimentation he does not run the risk as in our culture of associating this wickedness with sex itself rather than with sex at his age. The adult in our culture has often failed to unlearn the wickedness or the dangerousness of sex, a lesson which was impressed upon him strongly in his most formative years."[2] This commentary on the origins of sexual disturbances throws out a real challenge to conventional psychoanalytic theories dealing with the same problem.

Benedict goes on to observe that some primitive societies do make discontinuous demands but that resultant strains are experienced collectively through the "solid phalanx of age mates" and supported by graduation into traditionally prestigeful secret societies with elaborate initiation rituals interdicting previous behavior and sanctifying new role expectations. Our own difficulties then appear in new light: "It is clear that if we were to look at our own arrangements as an outsider, we should infer directly from our family institutions and habits of child training that many individuals would not 'put off childish things'; we should have to say that our adult activity demands traits that are interdicted in children, and that far from

---

[2] Benedict, Ruth, in *A Study of Interpersonal Relations*, edited by Patrick Mullahy, Hermitage Press, Inc., New York, 1949, p. 305.

redoubling efforts to help bridge this gap, adults in our culture put all the blame on the child when he fails to manifest spontaneously the new behavior or, overstepping the mark, manifests it with untoward belligerence. It is not surprising that in such a society many individuals fear to use behavior which has up to that time been under a ban and trust instead, though at great psychic cost, to attitudes that have been exercised with approval during their formative years."[3] So this seemingly simple thesis conceals a double criticism of a culture which both creates immense discontinuities in vital areas of life and ignores the necessity for providing transition rituals when major strains arise. This double failure leaves the members of such a society eternally tied to their childhood regardless of chronological aging, even as the necessities of adult role-playing entail suppression of these childish "fantasies" lest they interfere with adult tasks. There are two main outcomes of this double failure: insufficient suppression of fantasy can lead to inability to distinguish it from reality and consequent mental breakdown. Overly firm suppression shrivels affective life to the point where the enjoyment of spontaneous fantasy becomes impossible. In either event, childish experience and experiential modes are not assimilated and therefore adult life remains significantly impoverished.

Along with Benedict's essay, far more comprehensive work has been done on the ways whereby primitive societies guarantee their members a full life. The whole corpus of Paul Radin's work is directed at this problem. His interpretation of primitive social structure and culture as summarized in *The World of Primitive Man* documents in great detail the defenses of individuation embedded in primitive social organization. It is unnecessary to trace his long dispute with the proponents of theories about the insufficiencies of "the primitive mind" except to point out that this dispute was part of his larger aim in that it entailed elucidating the exceptional balance be-

[3] *Ibid.*, p. 308.

tween reason and emotions which distinguish the socialized primitive from his civilized brethren. Here his convergence with Jung strengthened both their positions though no one has yet systematically explored the full implications of their common discoveries. Jung uncovered among his patients a twisted need for the same life syntheses that Radin found to be the cherished core of primitive social structure.

There is no point in trying to reproduce Radin's dense picture of this social structure. Doing so is as difficult as paraphrasing a poem and, indeed, doing so would entail paraphrasing the many poetic and philosophic expressions with which his work is studded as they point at essentials of primitive life. Who can condense the saga of Trickster without doing serious injustice to it? Jung, Kerenyi, and Radin himself do try in the volume *The Trickster* but the story itself remains, like all works of art, immune to final paraphrase. It is this literary quality in Radin's work, the acceptance of the finality of poetry and philosophy, that leads him to describe the primitive world-view by quoting examples with only slight analytic adumbrations and that accounts for both its depth and its elusiveness. He asks us, by exercising our imaginative faculties, to see through the primitive's eyes; and if we are not ready or able to do so, both Radin and the world view he strives to recapture remain beyond our reach.

But there is more of science in his work than this. He actually does provide an abstract conception of institutional functioning in primitive society which can stand with the best anthropological and sociological theorizing. His interpretation of primitive economics seems closer to the facts as ethnographers have reported them than many of the interpretations of these same ethnographers. If one accepts his central thesis that primitive economies exist to ensure each member of the society an irreducible minimum of adequate food, shelter, and clothing on the ground that—"Being alive signifies not only that blood is coursing through a man's body but that he ob-

tains the wherewithal to keep it coursing"⁴—then his interpretations of primitive ownership, exchange, and distribution
are indisputable. In this context, the intermixture of utilitarian
and non-utilitarian concerns in primitive economies can be explained in their own terms as products of a fundamental purpose utterly different from that underlying both our economy
and our economics. So, his seemingly slight shift in emphasis
when interpreting the flow of goods in primitive society clarifies a broad variety of specific manifestations:

"In general, the tendency has been to speak of all aspects
of primitive economics connected with transfer, barter and
purchase, as if their main function was to serve as an outlet
for the expression of specific human emotions and as if there
was not a rigorous restriction of purely personal activity in
such matters.

". . . What apparently Fortune, Malinowski and Thurnwald
seem to have failed properly to understand and stress is that
one of the primary roles, if indeed it is not actually the primary, of a transfer and exchange is to visualize, dramatize
and authenticate the existence of certain fixed relations sub-
sisting between specific people and that this relationship has
a 'monetary' value. The actual reaffirmation of this relationship
. may take an exceedingly short time and the non-material emol-
uments flowing from it a very long time. That is, after all, true
of every type of exchange and transfer. It is an unjustifiable
procedure to relegate the utilitarian aspect of a transfer among
primitive peoples to a secondary position because of the richness and the duration of its non-utilitarian accessories, just as
unjustifiable as it would be to do the same in our own civilization."⁵

By hewing to his conception of primitive life as balanced
between utility and ceremony rather than viewing it as being

⁴ Radin, Paul, *The World of Primitive Man*, Henry Schuman, New York,
1953, p. 106.
⁵ *Ibid.*, pp. 126, 130.

as lopsided in the direction of the latter as we are toward the former, he is able to comprehend the deeper "utilities" underlying property transfer and illuminate the brilliant ethnography contained in Malinowski's description of the "Kula" without imposing alien categories on it.

Radin's analysis of primitive economics contains a highly original interpretation of the relations between economics, magic, and religion. He takes account of the exploitative manipulation of magic and religion when the fundamental purpose of the primitive economy has been perverted by the emergence of a surplus and ruling class. Underlying this is a vital distinction between socially creative ceremonial magic which enhances living and socially destructive magic which renders men subject to unnecessary exploitations.

With this distinction in mind, Radin's treatment of the crises of life and their rituals is immensely revealing. He starts by recognizing, as does Benedict, that growth from childhood to adulthood entails a major transition at adolescence. He also notes that resolution of the strains accompanying this transition is always accomplished in such a fashion as to preserve the vested interests of the older people. This "subjection and domination" is only transformed into exploitation when the benefits are reaped and accumulated by individuals as such rather than consumed immediately and ceremonially so that the persons enjoying them sanctify the rights of all to their irreducible minimum and to their private life cycle. It is hard to keep attention focused on this vital distinction and Radin does not always help as much as he might have. Perhaps it is so clear to him that he never understood the difficulty others might have with the distinction. One almost has the feeling here that on these matters Radin is so sympathetic with the movement of the primitive mind as almost to forget how remote this is from the spontaneous movement of civilized minds.

For that reason perhaps, useful as his chapter on the crises of life and their rituals may be, too much is left for the untutored imagination. Here, we need the help of other students like Stanley Diamond and Meyer Fortes to clarify the mysteries of the ritual dramas into which Paul Radin inducts us. But even here his first words, though not the last, remain indispensable:

". . . puberty became not simply the recognition that an individual had reached the age of sexual maturity; it became dramatized as the period of transition par excellence: the passing of an individual from the position of being an economic liability to that of an economic and social asset.

"Two distinct sets of circumstances, one physiological, the other economic-social, thus conspired to make of puberty an outstanding focus which was to serve as the prototype for all other periods interpreted as transitional. It was certified and authenticated by magic and subsequently sanctified and sacramentalized by religion. Its social and economic significance and evaluation are attested by the fact that the simplest tribes, the food-gatherers and fishing-hunting peoples, have already developed intricate and complex initiation rites around it. These puberty rites are the fundamental and basic rites of mankind. They have been reorganized, remodelled and reinterpreted myriads of times and of their analogy, have been created not only new types of societal units, such as secret societies, but new ideological systems as well."[6]

He goes on to place the puberty rites at the central and vivifying focal point from which rites and observances celebrating birth and death radiated so that the phases of human life were knit together, perhaps as the poet would have it, "bound to each other by natural piety." Separation and reintegration were accomplished through the social biological formula involving death in one status with subsequent rebirth in another on a higher level. Underlying all of the great ritual dramas is the impulse: ". . . on the one hand, to validate the reality of the physical, outward world and the psychical inward

[6] *Ibid.*, p. 152.

world and, on the other, to dramatize the struggle for integration, that of the individual, the group and the external world. This is done in terms of a special symbolism which is expressed in actions and in words, a symbolism which represents the merging of images coming from within and from without. This validation, finally, is articulated artistically and creatively by individuals peculiarly qualified, emotionally and intellectually, that is, by the thinker and the religious formulator."[7] So the old prestidigitator leads us to the periphery of the mystery by juggling inner and outer symbols, leaving those of us who are at least as peculiarly qualified as primitive "men of action" to pick them up when he drops them himself. That this is no easy task, all who have tried will testify.

Understanding these ritual dramas in the context of the part they play in protecting the primitive life cycle can be enhanced by closer inspection of the social and psychic mechanisms through which their effects are achieved. In order to show that anthropologists are still thinking about this problem, some comments by Meyer Fortes will be quoted at length:

"Primitive people express the elementary emotions we describe by terms like fear and anger, love and hate, joy and grief in words and acts that are easily recognizable by us. Some anthropologists say that many non-European peoples are sensitive to the feeling of shame but not to guilt feelings. I doubt this. One of the most important functions of ritual in all societies is to provide a legitimate means of attributing guilt for one's sins and crimes to other persons or outside powers. In many primitive societies this function of ritual customs is prominent and it leads to the impression that individuals have a feeble sense of guilt, by comparison with Europeans. The truth is that our social system throws a hard and perhaps excessive burden of moral decision on the individual who has no such outlets for guilt feelings as are found in simpler societies. This is correlated with the fragmentation of social rela

[7] *Ibid.*, p. 172.

tions, and the division of allegiances and affectations in our society. I am sure it has a great deal to do with the terrifying toll of mental disease and psychoneurosis in modern industrial countries. We know very little about mental diseases in primitive communities. What evidence there is suggests that those regarded by many authorities as of constitutional origin occur in the same forms as with us. But disturbances of personality and character similar to those that cause mental conflict and social maladjustment in our society seem to be rare. I do not mean to imply that everybody is always happy, contented, and free of care in a primitive society. On the contrary, there is plenty of evidence that among them, as with us, affability may conceal hatred and jealousy, friendliness and devotion enjoined by law and morals may mask enmity, exemplary citizenship may be a way of compensating for frustation and fears. *The important thing is that in primitive societies there are customary methods of dealing with these common human problems of emotional adjustment by which they are externalized, publicly accepted, and given treatment in terms of ritual beliefs; society takes over the burden which, with us, falls entirely on the individual.* Restored to the esteem of his fellows he is able to take up with ease the routine of existence which was thrown temporarily off its course by an emotional upheaval. Behavior that would be the maddest of fantasies in the individual, or even the worst of vices, becomes tolerable and sane, in his society, if it is transformed into custom and woven into the outward and visible fabric of a community's social life. This is easy in primitive societies where the boundary between the inner world of the self and the outer world of the community marks their line of fusion rather than of separation. Lest this may sound like a metaphysical lapse, I want to remind you that it springs from a very tangible and characteristic feature of primitive social structure, the widely extended network of kinship. The individual's identification with his immediate family is thus extended outward into the greater society, not broken off at the threshold of his home."[8]

This passage is quoted in full because its profound comparison of primitive and civilized situations points toward a breakdown of the boundaries between sociology, anthropology, and psychiatry along lines designed to ensure their collective reconstitution as a genuine science of man in which the problems of healthy and pathological human functioning will emerge as central. Because of the dependence of ritual drama on the state of the arts, the humanities would necessarily play an important role in filling out this reconstituted science of man. As one indication of this, Francis Fergusson in *The Idea of A Theatre*, working in the tradition of Jane Harrison, develops an interpretation of drama that beautifully complements the anthropological contributions. His interpretation of *Oedipus Rex* as ritual encompasses the insights of Freud but avoids psychoanalytic reduction, by showing how the play functioned in the context of its place within the Greek Festival of Dionysos, where it acted on the developed "histrionic sensibilities" of the audience. They were able to experience the rhythms and impulses behind tragic action without actually undergoing its disasters.

Paul Radin's synthetic view of primitive institutions includes a detailed interpretation of their political-legal structure and patterns of personal and social status. Primitive government consists of extended kin organizations like the clans with tribal chiefs serving to symbolize authority, while clan leaders actually wield authority over their kin when necessary. Custom hedges in persons wielding authority as much or more than it does their subordinates. Legal codes rarely appear until the kin authority system begins to give way to state forms. These are usually imposed by emerging national authorities and opposed by the tribal system. According to Radin, all important

[8] Fortes, Meyer, in *The Institutions of Primitive Society*, a series of broadcast talks by Evans-Pritchard, et al., The Free Press, Glencoe, Illinois, 1956, pp. 89-90, (emphasis ours, M.R.S.).

statuses in primitive society are kin statuses so that marriage becomes essentially the uniting of two kin groups. Individuality is status so that status ceremonies become vehicles whereby the primitive expresses his individuality at the same time that he reaffirms his social existence and the social existences of his relatives.

Primitive social order as conceived by Radin rests on a dramatic synthesis in which everyday life is imaginatively transformed and saturated with meaning. Individuality depends on the capacity to participate imaginatively in the experiences and satisfactions of the whole community. Men and women, old and young, weak and powerful, all have their place in the tribal order so that status guarantees the privilege of participation as well as assurance of the irreducible minimum required to sustain it. Western thought patterns interfere with sympathetic appreciation of this kind of individuation. In a sense, it is the deterioration of this same imaginative faculty so central to primitive life that renders us incapable of apprehending the fashion in which it releases human potentialities. In an essay on aesthetics, the British social anthropologist E. R. Leach, comments on the symbolic powers of primitives expressed in daily life:

"Whereas we are trained to think scientifically, many primitive peoples are trained to think poetically. Because we are literate, we tend to credit words with exact meanings—dictionary meanings. Our whole education is designed to make language a precise scientific instrument. The ordinary speech of an educated man is expected to conform to the canons of prose rather than of poetry; ambiguity of statement is deplored. But in primitive society the reverse may be the case; a faculty for making and understanding ambiguous statements may even be cultivated.

"In many parts of Asia, for example, we find variants of a courtship game the essence of which is that the young man first recites a verse of poetry which is formally innocent but amorous by innuendo. The girl must then reply with another poem which matches the first not only in its overt theme, but also on its erotic covert meaning. People who use language in this way become highly adept at understanding symbolic statements. This applies not only to words but also to the motifs and arrangements of material designs. For us Europeans a good deal of primitive art has a kind of surrealist quality. We feel that it contains a symbolic statement, but we have no idea what the symbols mean. We ought not to infer from this that the primitive artist is intentionally obscure. He is addressing an audience which is much more practised than we are at understanding poetic statement."[9]

It is this trained imagination that allows the primitive to participate in the status dramas of daily life. Perhaps it even allows him to dramatize the natural world so as to see in it regularly what only our painters and poets can see and they only sporadically.

This is a long distance from Levy Bruhl's conception of "participation mystique." In no sense does imaginative symbolization preclude logical thought or separation of self from the symbols or the objects symbolized. Our tendency so to see it is probably a cultural reflex arising from the deep split between logic and emotion in our own daily lives. Primitive life entails no such split so that duty, will, and impulse are imaginatively apprehended rather than explained logically— and imaginatively apprehended in a manner that admits the claims of all three even as the requirements of status are fulfilled. Here is where Radin's contribution to our understanding of primitive life is most important as well as least susceptible to paraphrase. *Primitive Man as Philosopher* reveals abundantly the complex perceptions of human nature and the human condition embodied in primitive folklore, mythology, and religion. These perceptions, however, like the "obscure" character of primitive art referred to by Leach, require a sym-

[9] Leach, E.R., *ibid.*, pp. 29-30.

bolically imaginative response for comprehension. Anything less is likely to convert them to "parables" or explain them in one or another reductive frame of reference.

When Radin tells us that the Winnebago narrative about Trickster, which he so carefully preserved, contains a profound moral regarding the dangers of instinctual, non-socialized behavior, he is providing us with a clue for grasping the meaning of the story. The point here is that no Winnebago would ever have needed any such clue. The meaning of the story was directly and symbolically apprehended and its relevance to his imaginative reconstruction of his own experiences clarified as he assimilated the symbols and symbolic events in the story. Here is art functioning to modify consciousness directly, as it only occasionally does for us. Until we are able to feel its rhythms, even Radin's sensitive interpretation leaves us outside the story. Only when we let it infect our inner life—that is, experience it imaginatively—can we see how it could affect behavior. For we have to feel the Trickster in ourselves as the primitive does quite spontaneously before we can appreciate its tragic consequences or conquer it. Being told that the narrative ridicules Trickster's blind striving can hardly take the place of the wisdom that comes from sensing the folly in unguided instincts, both our own and those of our fellow men, any more than maturity can come from memorizing theories about human growth. The wonderful Winnebago philosophic tale, *The Seer*, reproduced in *The World of Primitive Man*, tells us more about human limits and the tragedies of human over-reaching than many books on ethics. But it speaks first to the ear of the heart and only when that ear listens can the ear of the head hear what is being said.

The combination of artist and philosopher in the role of the primitive thinker as distinct from the man of action is not as removed from civilized actuality as many would contend. Poetry and philosophy are intimately interrelated as diverse figures like George Santayana, T.S. Eliot, and Wallace Stevens have argued and exemplified. But the conception that these activities *must* be interrelated is alien to our specialized civilization. And even more alien is the relation between primitive thinkers and men of action which rests upon the thinker's ability to sense crises of the community and cope with resulting strains by symbolic and ceremonial acts. While men of action live in a "blaze of reality," there are strains in their relation to their impulses, to the community, and to the external world. Thinkers who perform properly feel these strains first and express them symbolically. Religious men, shamans and priests, cooperate in this endeavor and indeed are occasionally themselves the artist-philosophers of the tribe. Radin's complex interpretation of primitive religion denies the theories of "mystical participation" without denying that the bulk of primitives who are non-religious still have their experience illuminated by their relation to the authentic religious men of the tribe. Actually there is always a possibility that the tribal intellectuals will become exploitative, but the larger context of tribal status should keep this tendency within limits.

In terms of a perspective on the modern community, the distinctions between men of action and thinkers or between religious and non-religious men must be seen as entailing important points of contact and even fusion between the distinctive groups. Primitive artist-philosophers articulate the symbolic-ceremonial web of the tribe while religious men authenticate this web by inspiration and the evidence of their "seizures." Both are more sensitive to strains and tensions than ordinary members of the community and in their different ways both react to these strains in order to cushion their impacts on the less introspective members. But all remain tied to each other in the larger network of kin statuses and the experiences are shared insofar as they can be symbolically communicated. The revelations of the shaman are the property of the tribe.

Unfortunately modern counterparts of these primitive creators are hard-put to find a similar context for their own activities though many of them do indeed search for it. The turn toward magical doctrine and Celtic fables by a great poet like Yeats is one such manifestation. But the modern artist, mystic, or philosopher rarely breaks through to community experience, nor does he help to authenticate communal symbols. Modern men of thought are segregated from the everyday world and the people who live in it by barriers of sensibility and language. Our artists are therefore forced to record their private responses to the strains of civilization without any assurance that the meaning of their expression will carry much beyond a small circle of similarly inclined creators and critics.

Impersonal social agencies, the schools, churches, city halls, and newspapers of the modern city assume perfunctory responsibility for celebrating the major life transitions. Indeed, these life transitions are often passed over in comparative silence. Since the adults have to a large extent lost control over their changing communities, they are neither inclined nor equipped to initiate succeeding generations. On the face of things, more distress is aroused over threats to job security than over the assumption of masculine or feminine duties. Social roles and role transitions become occasions for anxiety rather than vehicles for human fulfillment. There is an unfortunate inner dynamic in modern "spurious" community life wherein the very spuriousness creates anxiety which propels the climb to new levels of status in the hope that the gnawing will cease, yet this upward movement only leads to further anxiety aroused by the insufficiencies at the new level. The people involved soon lose their capacity for distinguishing status anxiety connected with important life transitions from the myriad status threats that daily life presents. There are few communal rituals to help discriminate the real from the trivial nor can relatives provide much guidance through stormy passages. The intellect is a weak crutch, since the complexity

[247]

of modern existence demands broader perspectives than most people can muster while emotions are even more unreliable so long as childish impulses reign.

It almost seems as if community in the anthropological sense is necessary before human maturity or individuation can be achieved, while this same maturity is, in turn, a prerequisite for community. This is an oversimplification since we know that some people achieve integrity in spite of frustrating community experiences. The real problems and transitions of life have a way of breaking through even where appropriate ritual occasions are absent and some people manage to lead fairly genuine lives in a spurious culture. Sociologists must search out such people to study the conditions that made their achievement possible. This could lead to further understanding of community patterns arising within our complicated civilization with potentialities for releasing true individuality.

There is little to be gained by sentimentalizing about primitive life or advocating a return to it in any form. We are far too deeply committed to urban-industrial civilization even to think of abandoning it now. Nor can we artificially incorporate outposts of "folk culture" within our own context since they quickly deteriorate into "pseudo-folk" forms when ripped from their proper setting. Folk music written on Tin Pan Alley or jazz produced by classical musicians, whatever its intrinsic merits, is hardly a satisfactory solution. Instead, the image of primitive society supplied us by Radin, Redfield, and Sapir, in which integral human functioning through an intelligible life cycle where major human needs are assured of satisfaction and major life transitions directly confronted, helps us to formulate norms for human community life.

Anthropologists also provide a good many specific clues about the circumstances of life in a genuine culture even though these circumstances cannot be directly reproduced in a civilized community.[10] Redfield's most recent book, *Peasant*

[10] There is another group of anthropologists whose work seems directly

Society and Culture, deals with peasant communities which he regards as an intermediate type falling between folk or primitive society, on the one hand, and urban society, on the other. He develops theoretical models for studying the linkages between urban and peasant social patterns coexisting within the same national framework. Sociologists are familiar with this problem. Most of the sub-cultural diversity of which Park and the Chicago sociologists during the twenties were so fond stemmed from the presence in the city of first-generation immigrants, many of whom retained peasant values and social patterns. Redfield's sensitive distinction between the great or sophisticated-urban tradition and the little or peasant tradition throws light on the familiar division between high and low culture.

Community sociologists would be well advised to ask themselves why the most important book on methods of community study in recent years should have come from an anthropologist. This is, of course, Redfield's exemplary handbook, *The Little Community*, which contains far more than technical aids. It presents a number of alternative complementary conceptual frameworks as well as a highly sophisticated philosophy of research. It is as applicable to the study of cities or urban sub-communities as to peasant or folk societies so that the outworn distinction between anthropological and sociological field methods is here exploded in the most convincing fashion possible.

---

relevant to the theory of communities presented in this book. That is the group around Julian Steward. His monograph on area research and his book on culture change emphasize the need for an evolutionary approach to communities that seems quite compatible with the theoretical position adopted here. I might also mention Conrad Arensberg's parallel effort at developing a functional theory of communities in an evolutionary context. I discovered Steward's and Arensberg's work after my own formulations had been developed. It was too late to try to deal with their relationships to this theory, but there are important points of convergence. See especially Steward's SSRC monograph *Area Research*, and Arensberg's essay "American Communities," in the *American Anthropologist*, Vol. 57, 1955, pp. 1143-1162.

Even more striking, however, is Redfield's ability at combining methods for studying social structures from outside with techniques for exploring the inner perspectives of participants so that his battery of approaches ranges from ecology to life histories and the delineation of typical world outlooks. Here the "insight" of the humanist is combined with the "objectivity" of the scientist to present a richer conception of social science than one confined to either alternative.

Sociologists then can fruitfully turn to contemporary anthropology for perspectives on all phases of community life. Our whole interpretation of the structure of folk society on which so many of our concepts rest must be modified according to the theories and findings developed by Radin, Sapir, Redfield, and the British anthropologists quoted earlier. These concepts and findings acquaint us with unique community systems in which important human potentialities are fulfilled in a fashion peculiarly alien to present-day America. It is exactly our collective commitment to ever-receding status goals that makes the contrasts provided by primitive peoples unusually apt. By cutting through the confusion between "success" defined in marketing terms and the achievement of integrity as a human being, these studies shed light on a dark aspect of American life.

PSYCHOANALYTIC PERSPECTIVES ON THE
MODERN COMMUNITY

PSYCHOANALYTIC concepts and therapies have become important elements in the daily life of some urban sub-communities. The exurbanites, for example, are all too familiar with both. Many of them have been patients, others have experienced arduous readjustments when members of their family underwent treatment, and all are regularly exposed to heavy doses of the psychoanalytic jargon which suffuses the communications industries. Many of the ubiquitous experts in Crestwood Heights are "analytically-oriented." So too are the case workers in Cornerville. And the inhabitants of Greenwich Village have a broader range of choice among psychoanalytic schools than they do among breakfast cereals. This all converges with a growing awareness of mental illness on the part of the general public. Mental illness promises to replace polio as the medical fund-raiser's magic disease. Supplementing this, the mass media have turned to psychiatric themes. The psychiatrist threatens to become as familiar and as stereotyped a figure as the old family doctor. Tranquilizers used by mental patients and the general public alike form a bridge between anxious people inside and outside hospitals. Increasing recognition and hospitalization of the mentally ill today means that more families have to face the complicated problems occasioned by the institutionalization of a close relative.

There are indications that community studies over the next few years will focus increasingly on the problem of mental health. *Crestwood Heights* is one volume in a series on this

theme. Other studies in progress or completed by skillful social scientists like Alexander Leighton, Thomas Rennie, Leo
Srole, August Hollingshead, Frederick Redlich, Edward Linde-
mann, Florence Kluckhohn, John Spiegel, John Clausen, and
others suggest that an excellent literature will soon be available
on this aspect of community life. Most of this research is as
yet unpublished, so that efforts at fitting it into a general
theory of community development must wait. Perhaps this
research focus will yield the same kind of comprehensive data
that the studies of delinquency in the twenties and thirties
provided. One hopes that these investigators will profit from
the lessons of the delinquency studies by orienting their research around the effects of changing social structures in
specific sub-communities as these changes are caused by the
manifestations of urbanization, industrialization, and bureaucratization in the fifties. Faris and Dunham in *Mental Disorders
in Urban Areas* point in this direction, though their acceptance
of ecological concepts as explanatory rather than descriptive
weakens their analysis. However, the search for relationships
between sub-community constellations and mental disorders
shows promise as long as close attention is paid to the specific
interconnections.

All of this suggests that mental illness may be a central
symptom of community disorganization in the fifties. This does
not mean, of course, that delinquency has been eliminated.
Indeed, it has actually diffused upward so that today we find
middle-class adolescents adopting such delinquent patterns as
gang violence. To explain this diffusion is one of the main
tasks confronting community sociologists. But the shift to mental illness as a focal symptom implies that in the present period
it embodies more central dislocations than delinquency. It is
worth noting that both delinquency and mental illness are
closely linked to family disorganization. The family conflicts
of the twenties provided a context for delinquency while the
family problems of the fifties play the same role in regard

to mental illness. Before satisfactory interpretations can be advanced, it is necessary in both instances to look beyond the family to the entire institutional network with which it is bound up.

Careful reading of the psychoanalytic literature shows that Freud anticipated most of the main problems and directions of research later seized upon and expanded into full-blown alternative theories by former disciples. But this inclusiveness does not preclude the possibility that the narrower focus of the dissidents or the neo-Freudians will not yield sharp insights of great importance. So the careful exploration of dimensions of personality malfunctioning offered by Adler, Jung, Rank, Fromm, Horney, and Sullivan, to mention but a few, cannot be ignored because they are less inclusive than Freud. Their work has to be placed in proper context and recent efforts at doing so like Ruth Munroe's *Schools of Psychoanalytic Thought*, and Ira Progoff's *Death and Rebirth of Psychology* prove that this task can be accomplished. From inside the orthodox camp, essays like those published in the volumes of *Psychoanalytic Studies of the Child* by Heinz Hartmann, Ernst Kris, Rudolf Lowenstein, Edith Jacobson, and others show, under the guise (or disguise) of "ego psychology," increasing willingness to incorporate the insights of the deviators. Actually this process of incorporation serves to broaden the conceptions of the original theory and there is no reason to assume that equal progress could have been made without the sharper emphases of the dissidents.

Freud's influence on community sociology is not obvious. Except for instances like Dollard's *Caste and Class in a Southern Town*, where the author had analytic training, or the analysis of James West's material from Plainville by Abram Kardiner in *Psychological Frontiers of Society*, it has to be searched for carefully. At the highest level of theorizing, Park was undoubtedly familiar with Freud. His conception of the city as a place where human impulses of all kinds can come to expression is

[253]

sobered by awareness that some of these impulses may well
be destructive to social order. Park's analysis of collective behavior, especially his interpretation of mass phenomena, sees
the place of unconscious impulses clearly and in a manner not
unlike Freud's. But neither Park nor his students succeeded in
incorporating Freudian concepts into their empirical research
on any large scale.

This remains yet to be accomplished by later students of
communities. It has to be done simply because we now know
that many institutional processes have unconscious underpinnings which affect social life significantly. Dollard's study of
Southerntown makes clear some of the ways in which psychoanalytic insight clarifies social patterns. Thus Southern White
commitment to the caste system in face of Northern pressures
as well as the inner conflicts it creates can hardly be understood unless one studies the neurotic secondary gains that it
provides. So long as their whole system of sexual and aggressive satisfactions depends on caste prerogatives, efforts at
change will always meet determined resistance and possible
violence.

Though it is impossible to summarize all of the directives
for community research implicit in the vast body of Freudian
hypotheses and therapeutic strategies, the one that strikes us
as perhaps most important is simply the recognition of unconscious motivation itself. This sensitizes the sociologist to
the fact that the people he is studying may unwittingly "explain" their own behavior in terms which conceal unacceptable
motives, especially those connected with forbidden sexual and
aggressive impulses. Freud provides tools for exploring these
unacceptable impulses as they may find expression through
dreams, slips, and even the art forms of the community.

Yet, so new and perhaps disturbing is this conception that
even the general orientation toward sexuality has not been
seriously included in American community studies. Thus,
Crestwood Heights, informed by psychiatric theory as it is,

[254]

tells us almost nothing about the sex lives of the suburbanites. Hints are thrown out that it is a source of difficulty but they are never followed up. Perhaps this kind of data was unavailable to the authors but this seems unlikely when one remembers that they were working through a mental health clinic. Apparently it was their own resistances, rationalized perhaps as discretion, that led them to gloss over this aspect of life. Even after Kinsey's courageous statistical survey, we know more about the sex life of the savages than of suburbanites.

Freud is concerned with adult sex life in its developmental context. Here *Crestwood Heights* provides more clues but still not enough. We know that the children are taught "healthy" attitudes toward their bodies but this aspect of socialization is only sketchily presented. Perhaps later volumes in the series will have more to say. Or possibly this kind of information simply could not be obtained without endangering rapport with the community. In any event, it is clearly an essential order of data and sociologists will have to devise methods, possibly of a projective nature, to get at it if they hope fully to explore a modern community. By calling attention to the bodily impulses and their training, Freud forces us to attend to an aspect of community life which is ordinarily overlooked in American communities by American sociologists. If nothing else, our standards of ethnography have to be broadened accordingly.

However, Freudian conceptions can help even more than this. By providing a frame of reference for interpreting rigidities in the symbolic systems of community members, they allow insight into degrees and kinds of self-alienation experienced. This does not refer to the kinds of formulations usually advanced by "culture-personality" theorists. It dips into the collective psyche at a more general level to ask the question: "How much freedom is there among the adults of a given community for symbolizing new experience through preconscious functioning?" While this would obviously be related to their early socialization experiences, opportunities offered by institutionally patterned adult roles are equally important.

This is another way of saying that sociologists must attend to the emotional life of the people they are studying; especially to the extent to which that emotional life is dominated by compulsions and conventional categorizations which interfere with appreciating complex human and natural realities. The assumption underlying this interpretation of Freud is that mature human functioning rests upon a creative capacity to modify self-symbolizations and symbolization of others in the direction of increasing comprehension of novelty and complexity. This, in turn, depends upon willingness to confront inner conflicts as they manifest themselves in present behavior. It presupposes a life cycle in which the individual is truly socialized toward acceptance rather than repression of his own impulse life, and a recognition that such acceptance does not mean translation into action but rather integration into the system of larger personal purposes.

In this view, the therapeutic process is a way of dissolving unconsciously·supported rigid self-conceptions and conceptions of others. Analysis of transference is the main instrumentality insofar as it involves careful exploration of the affects the patient "transfers" onto the neutral analyst. Dream analysis, free association, and other technical tools help to elucidate the meaning of these transferences in terms of repressed traumatic experiences and developmental fixations. But the purpose of therapy remains the unfreezing of frozen affects to enrich the patient's inner life and his awareness of objective opportunities. It helps him to develop an identity based on assimilation of the conflicts in his own life experiences so that he can respond to others and their problems without distortions imposed by his own unresolved conflicts.

This is done by "working through" these same life experiences with the therapist so that impulses which the child was forced to repress can be reexperienced by the adult and put in proper perspective. Attention has to be paid to fantasies and dreams. The therapist dignifies the patient's present feelings by attending to them closely and establishes the relevance of his past experiences by encouraging him to recall and re-experience them as fully as possible. As has often been noted, this therapeutic relationship allows for full human expression of a kind rarely tolerated outside of it. It is true that it also involves restrictions, but the range of expressivity allowed the patient is remarkable in our culture. He is the focus of attention and the events of his life are the subject of joint study. In this sense, it almost resembles a tribal ritual and indeed the analogy could be pushed much further in that it is also the occasion for the symbolic death of an old personality followed by the rebirth of a fuller one.

Freud perceived the mythology of every man's life, and the analytic relationship provides perhaps the only place where contemporary men and women can explore the drama of their own lives. It is set apart from other reality-bound enterprises and deliberately creates a fantastic atmosphere to sustain the exploration. Here the conventional modes of communication are set aside so that suppressed faculties can come into play. The patient becomes an artist participating in the identical self-creative process that is one phase of the aesthetic act.

However, as the chapter on anthropology has shown, some aboriginal societies do not depend on the highly precarious commitment to a transient therapeutic relationship to help the individual confront his own destiny. Ritual dramas bind together the phases of life to assure growth. It is important that the sociologist search for the sources of symbolic flexibility, or ritual death and rebirth, in our own society. Freud has placed this search on a sound basis by describing some of the major crises of the child and telling us about some of the patterns of repression resulting from them. Furthermore, he has given society one set of techniques for restoring equanimity to a

[257]

privileged few and he has given social science leads for exploring the conditions under which these techniques can be rendered influential outside the clinic and the hospital.

The problem, then, is to assess the provisions which a community makes for meaningful life experiences in the context of a meaningful life cycle. This is the human purpose of any community study and it must underlie any theory of disorganization. Freud deepens our insight into the psychic mechanisms whereby this is to be accomplished. Later psychoanalysts may provide us with concepts that fit contemporary experience more satisfactorily, but the central insight as to the place of unconscious impulses and their binding role when they are stripped of contact with preconscious-conscious faculties as well as their liberating role when such contact is securely established remains Freud's contribution.

An important essay that views Freud's contribution in a similar fashion is Ernest Schachtel's *On Memory and Childhood Amnesia*. Aside from its demonstration that childhood amnesia is at least in part a cultural product, the whole essay emphasizes the need for continuity between the capacities of the child and those of the adult. He seizes upon the richness and spontaneity of children's perception as a faculty to be cherished and assimilated by the adult on a higher level and deplores the narrow schematization of experience in our own society: "Cultures vary in the degree to which they impose clichés on experience and memory. The more a society develops in the direction of mass conformism, whether such development be achieved by a totalitarian pattern or within a democratic framework by means of the employment market, education, the patterns of social life, advertising, press, radio, movies, best-sellers, and so on, the more stringent becomes the rule of the conventional experience and memory schemata in the lives of the members of that society. In the history of the last hundred years of Western civilization the conventional schematization of experience and memory has become increasingly prevalent at an accelerating pace."[1]

The reflection of this at an adult level is the impoverishment of linguistic imagination, precisely the faculty that poets possess to the highest degree. The child does not need it since his capacity to verbalize his perceptions is as yet unformed, but the adult, for whom words intervene between himself and the world, is lost without it: ". . . Experience increasingly assumes the form of the cliché under which it will be recalled because this cliché is what conventionally is remembered by others. This is not the remembered situation itself, but the words which are customarily used to indicate this situation and the reactions which it is supposed to evoke. . . . There are people who experience a party, a visit to the movies, a play, a concert, a trip in the very words in which they are going to tell their friends about it; in fact, quite often, they anticipate such experience in these words. The experience is predigested, as it were, even before they have tasted of it. Like the unfortunate Midas, whose touch turned everything into gold so that he could not eat or drink, these people turn the potential nourishment of the anticipated experience into the sterile currency of the conventional phrase which exhausts their experience because they have seen, heard, felt nothing but this phrase with which later they will report to their friends the 'exciting time' they have had. . . . But while Midas suffered tortures of starvation, the people under whose eyes every experience turns into a barren cliché do not know that they starve. Their starvation manifests itself merely in boredom or in restless activity and incapacity for any real enjoyment."[2]

Schachtel interprets this super-conventionalization as Freud does the super-repression of civilization in terms of a presumed

[1] Schachtel, Ernest G., in *A Study of Interpersonal Relations*, edited by Patrick Mullahy, Hermitage Press, Inc., New York, 1949, p. 45.
[2] *Ibid.*, pp. 13-14.

disruptive potential of the childish sensuous-emotional system. He observes, however, that this is especially true in a society like ours which channels all energies into the quest for status. Experience is reduced to its verbal expression and transformed into a status-counter before all else. One suspects that this is as true of the special vocabularies of "high culture" as it is of the advertising-spawned clichés of the less educated. Ten-letter words can be as effective blinders to experience as their four-letter counterparts.

Harry Stack Sullivan provides a theory that is peculiarly well suited for interpreting the rigidities in symbolic-perceptual functioning that characterize contemporary man. He starts with the problem of interpersonal communication and orients his theorizing toward explaining prevalent distortions in this process. Where Freud concentrated his attention on intrapersonal processes, Sullivan focuses on interpersonal processes. Neither of these writers was really as exclusive as their critics would suggest, so that the areas of convergence are actually considerable. Since this convergence cannot be demonstrated here in any detail, it will be assumed as a working hypothesis and an effort will be made to show how Sullivan supplies insights that build on Freud's to deepen a psychoanalytic approach to communities.

For Sullivan, as for Freud in his later writings, anxiety is the central human affect governing distortions in human relations. Objects, human or non-human, which evoke anxiety will not be perceived as they are. Instead, images bound to the perceiver's past will get in the way of his capacity to respond to a given object; his response will be governed by his past rather than by the object at hand. Anxiety evokes compulsive defenses, yet for Sullivan successful confronting of anxiety is the only basis for real growth. All of the defense mechanisms—repression (which he calls dissociation), reaction-formation, denial (which corresponds to his selective inattention), apathy, and the rest—break the communicative arc on which growth depends by fleeing from anxiety and the real object.

His developmental scheme highlights exactly this point. He identifies six major epochs: infancy, which lasts until the maturation of the capacity for language behavior; childhood, covering the maturation of the capacity for living with compeers; the juvenile era, extending to maturation of the capacity for intimacy with a person of the same sex; preadolescence, extending to the maturation of the capacity for intimacy with the opposite sex; early adolescence, extending to the patterning of lustful behavior; and late adolescence, in which a fully human repertory of interpersonal relations must be established if further maturing through adulthood is to proceed. Anything like full appreciation of this scheme requires careful examination of its detailed exposition in *The Interpersonal Theory of Psychiatry*. For our purposes, it is sufficient to note that truly individuated growth starts where the scheme ends. When the young adult has learned to relate without too much anxiety to individual members of his own sex and the other sex, he must then establish himself in work and familial roles of such a character as to permit an unfolding of his capacities *through* the relationships involved in these roles rather than in spite of them.

Adolescence then becomes a crucial phase. The Freudians tend to emphasize childhood, though they are beginning to recognize the importance of later epochs. Sullivan is concerned with showing that developmental fixations caused by bad experiences at earlier levels can be undone by good ones during preadolescence and later. Thus, the youngster who establishes a healthy friendship with another can learn a great deal from this about the range of interpersonal possibilities denied him in his previous experience. Sullivan devotes much effort to analyzing the agency of the personality that precludes such learning. He calls it the "self system," that portion of the psyche which is activated by anxiety to preserve threatened self-esteem.

Misunderstandings of this concept have led even friendly critics of Sullivan to assert that he denies the possibility of true individuality. Actually, his self system or reservoir of security operations does in our society often become the main meaning of individuality accessible to the person. Modern man experiences his selfhood as something apart from and opposed to the roles he has to play. This cherished selfhood is defended against contact with life because such contact would inevitably destroy the illusion of uniqueness on which it is based. To admit our community with others in the main realms of unpleasant emotional experience (anxiety, loneliness, insecurity, and the like) presupposes a capacity for sympathy as well as a willingness to admit "weakness," but the culturally proper man or woman in our society is one who suppresses these feelings in order to present an appearance of "successful competence." The self system organizes behavior so as to conceal the unpleasant affects, either confining them to private expression or eliminating them from consciousness altogether.

The image he presents to others remains the constant concern of the insecure individual. Since our society forces everyone toward insecurity in most relationships, the manipulation of "presented selves" becomes a major preoccupation. In part, this is what Riesman means by "other-direction." But Sullivan's conception probes more deeply since he assumes that this self-presentation involves anxious responses to threats perceived as emanating from someone else. This anxiety always carries with it distortions of the real situation. So the "other-directed" man aroused by status anxiety restores security by creating an imaginary "me-you pattern" in which his conception of both himself and the other person is seriously distorted. This is the psychopathology underlying the surface tendency that Riesman describes. It is not simply a matter of sensitivity to others. In addition, it involves an anxious response to another person

[262]

compressing him into a "manageable" image and treating him accordingly.

Thus Sullivan interprets the therapeutic relationship as one in which these rigid and anxiety-bound "me-you" configurations are dissolved, leaving the patient free to establish intimate relationships with others in which new self-other symbolizations emerge. Therapy proceeds by virtue of the therapist's capacity to observe the ebb and flow of anxiety in the patient and to follow up this observation by identifying the anxiety-arousing dynamisms. It entails exploring the developmental bases for "parataxic distortions" wherein the patient responds to the therapist as if he were a significant figure from the past. This is not too different from the analysis of the transference neurosis advocated by Freud, but Sullivan prefers to regard the therapist's task as "participant observation"—a strategy that throws emphasis on the therapist's self-observations insofar as these embody clues to the meanings of the patient's expressive behavior.

Without going further into the technical details of Sullivan's system, it is clear that he is concerned with the establishment of self-other conceptions which leave the individual open to novelty and growth. Stereotyped, anxiety-bound self personifications as well as personifications of significant others become the main impediments to the development of individuality. Developmentally, the stages of adolescence in which individuals learn to communicate with particular members of the same and opposite sexes are highlighted. If these phases are dominated by juvenile stereotypes, psychological adulthood cannot be achieved. Nor does Sullivan ignore the special difficulties involved in our culture. The maturing of the lust dynamism creates serious insecurities in a society in which sexual contact between adolescents is as circumscribed as it is in ours. Furthermore, our "taboo on tenderness" which becomes especially prominent in adolescence when impulse control is so necessary, leaves the intimacy need frustrated and renders its

[263]

integration with the lust dynamism and the security dynamism highly problematic.

It is precisely the virtue of Sullivan's theory that it focuses on the major difficulties in living that occur in a massively competitive society. These are the difficulties which lead to the formation of personalities dominated by the security operations of the self system at the expense of genuine individuation. People learn to go through the motions of communication without exposing their own depths or establishing any deep contact with others. Beneath the seemingly smooth surface where interaction is dominated by the exchange of clichés, parataxic distortions proceed unnoticed until their operation causes an outbreak of real disturbance. Sometimes this is occasioned by unanticipated life crises; at others, persons involved in a parataxic two-group are abandoned by their partners.

Our understanding of the difficulties confronting modern adolescents is greatly enhanced by looking at the work of still another psychoanalytic theorist, Erik Erikson. Treating a developmental phase dealt with extensively by Sullivan, i.e., adolescence, he formulates the problem in terms of the establishment of ego-identity:

". . . From a genetic point of view, then, the process-of identity formation emerges as an *evolving configuration*—a configuration which is gradually established by successive ego syntheses and resyntheses throughout childhood; it is a configuration gradually integrating *constitutional givens, idiosyncratic libidinal needs, favored capacities, significant identifications, effective defenses, successful sublimations, and consistent roles.*"[3]

". . . Man, to take his place in society must acquire a 'conflict-free,' habitual use of a dominant *faculty*, to be elaborated in an *occupation*; a limitless resource, a feedback, as it were, from the immediate *exercise* of this occupation, from the com-

[3] Erikson, Erik Homburger, "The Problem of Ego Identity," *Journal of the American Psychoanalytic Association*, Vol. IV, No. 1, p. 71, 1956.

panionship it provides, and from its *tradition*; and finally, an intelligible *theory* of the processes of life. . . ."[4]

". . . Whether or not a given adolescent's newly acquired capacities are drawn back into infantile conflict depends to a significant extent on the quality of the opportunities and rewards available to him in his peer clique, as well as on the more formal ways in which society at large invites a transition from social play to work experimentation, and from rituals of transit to final commitments: all of which must be based on an implicit mutual contract between the individual and society."[5]

Here we have a statement of the conditions necessary for minimizing the effects of the self system in order to give the adolescent an opportunity to find a place for himself. Erikson emphasizes that this can be accomplished only through a "moratorium," a period of experimentation, during which various identities can be tested. Exceptional individuals establish their own moratorium, and Erikson has shown how this worked in the case of George Bernard Shaw and Sigmund Freud. Most people, however, require a socially established moratorium.

Erikson argues that present-day adolescents are often denied any opportunity for consolidating differentiated identities. They suffer both from "identity diffusion," not being certain who they are, and from "identity foreclosure," the hasty espousal of self-conceptions to avoid the anxieties of diffusion without satisfactorily determining whether the identities so adopted are suitable for their unique capacities. In an earlier formulation, Erikson referred to the psychic condition of the American adolescent as one of "ego restriction" arising as a defense against the demands made by parents and society. Living in an utterly discontinuous culture, confronted by irrational demands from his parents, unaided by any significant rituals of transition, and deprived of adult roles which promise satisfactions in depth, affect is withdrawn from any of the alternatives and a life of quiet desperation accepted.

[4] *Ibid.*, p. 65.   [5] *Ibid.*, p. 73.

III. PERSPECTIVES

Locating responsibility for the establishment of adolescent identity in the community, this theory presupposes continuity between early "childish" identities and those established at adolescence. Looking forward, the individual should be able to see a series of later phases of adulthood each with its own demands and rewards. The human life cycle is a social as well as a biological fact. Only the most gifted and courageous are able to carve out their own. Unless the groundwork for identity elaboration is laid in adolescence, the adult is condemned to permanent confusion. A community must then provide its members with, at the least, meaningful sexual and work identities if it is to ensure its own continuity as well as the psychic integrity of its members.

Sullivan and Erikson present conceptions of the inner life of the modern normal-neurotic that are quite complementary. Sullivan shows a personality dominated by status anxiety unable to break through to real human communication, while Erikson isolates the psychic precondition for such a breakthrough in his concept of ego-identity. Before the individual can achieve syntaxic communication, he has to have a minimally secure sense of self. He must be able to let his parataxic distortions come to consciousness without allowing them to dominate his consciousness. He can do this only if he has worked through his past identities so as to establish their linkage with his present one. In other words, he must come to terms with his "childish" responses to the "family of the past" in such a fashion as to free him from their domination in the present. These responses or earlier identity fragments must be confronted and assimilated rather than repressed; otherwise their continued unconscious influence will be disastrous. With respect to sexual identity, the adolescent must learn that an automatic transfer of his attitudes toward his mother to girls and women of his acquaintance is as disastrous as similar transference of his needs and impulses toward his father onto other males. Where the society does not create conditions in which adolescents can learn about these differences and adapt themselves accordingly, the likelihood that fixed infantile patterns will dominate all relations with other people is maximized. If the society then provides a set of stereotyped "pseudo-identities" which the individual can easily if anxiously slip into, it is conceivable that social tasks can be accomplished even though the performers lack self-insight or empathy. Generalized social sensitivity, dominated by security operations, steer the individual through necessary contacts with others. The interaction is regulated by mutual avoidance of the unconscious overtones although these can occasionally seize control against the will of the participants.

To put the social psychiatric analysis a bit differently, Erikson views ego-identity as depending on the accessibility of roles in which acceptance by significant others is assured. Identity has to be confirmed by the environing community. However, when the significant others, both peer group and adult, have begun to question their own identities, this doubt is reflected back on the adolescent with subsequent anxiety and communication breakdown. His own doubts are enough to manage without shouldering the burdens of his parents and contemporaries. But it is exactly those burdens which Erikson finds decisive among adolescents suffering from pathological "identity diffusion." In the absence of identity models with which to experiment, the adolescent is compelled to adopt a set of role personalities compulsively. These allow him to participate with others in various areas of endeavor without really being committed or in communication.

Role personalities in modern urban society easily become objectified clichés. They congeal personal idiosyncrasies into networks of formalized expectations. When the pace of change is so rapid that the expectations themselves are no longer stable, then even the cliché identities that they are capable of engendering cannot be sustained. Thus the last source of identity, role personality, is shattered, leaving only a vague self-image of "flexibility" and powerful unconscious control by one's security system as guides through the social wasteland. It is true that the mass media provide some clues but these are even more abstract than the clues in the immediate environment. As a result, vital life choices and commitments are made_ haphazardly. The individual goes through the motions of gathering status-symbols in a life-space, bounded by dissociation and selective inattention. He never achieves the minimum security needed for questioning the activities into which he has stumbled, no less the worthwhileness of the status quest in the first place.

Psychoanalytic theories allow us to assess the psychic conditions for full participation in community roles. Starting with the integration of childish bodily impulses and images which is now seen as depending on the capacity of the parents to accept these impulses and body parts, the achievement of full adult functioning is traced through the whole life cycle. Later accomplishments, especially the attainment of ego-identity at adolescence, are highlighted in recent theorizing. Successful surmounting of these developmental levels depends upon the conscious-preconscious confronting of conflicts rather than fixation through compulsively repeated defenses. When developmental anxiety is repeatedly suppressed in the interest of security, emotional growth cannot take place. From the standpoint of community participation, sufficient emotional growth is required so that the individual can distinguish himself from others and can distinguish the people of the "present" from the images of the past. In other words, introjection and projection have to be held to a minimum while "transferences" must be worked through sufficiently to allow people to see others as they are rather than as "parental figures." *In this view, then, psychoanalytic therapy is a substitute mechanism for accomplishing developmental tasks that are necessary for helping human beings to reach full community participation.* Redefinition of the goals of therapy then usually implies implicit redefinition of the nature of community. When sexual malfunctioning was widespread, Freud developed his conception of genitality as the goal. Later theorists have been forced to reinterpret this goal along "interpersonal" lines so that the capacity for intimacy or conscious confrontation of anxiety and private fantasy come into focus. The malady that the psychoanalyst treats is always socially engendered and the cure always involves some form of restored social functioning.

In his early work, Freud focused on the way in which repressed infantile traumas led to neurotic symptoms. When some of the traumas reported by patients proved to have been entirely imaginary, he was forced to expand his etiological conceptions to include failures in achieving levels of libidinal growth and ego mastery as central predisposing causes for mental disorders. Rather than pressing for recall of specific traumatic events, he began to trace the development of characteristic responses and systems of responses in his patients. This shift was accompanied by increasing attention to "character resistances" which impeded therapy and finally led to recognition that character limitations were themselves a form of neurosis. Patients were coming for treatment with symptoms markedly different from the phobias and compulsions of the hysterical neurotic; their main complaints were boredom, diffuse dissatisfaction with life and apathy. In order to treat these patients, the analyst had to challenge the fundamental life orientation of the patient.

Freud saw that these character neuroses like the symptom neuroses had their origins in infantile experience. They consisted of defenses that may have been appropriate at earlier developmental levels but that quickly became liabilities when they persisted after the threats to which they were a response had disappeared. Freud's analysis of the pregenital and Oedipal origins of character defenses emphasizes the dominance of infantile fixations. Erikson and Sullivan turn their attention to the consolidation of these defenses during adolescence and young adulthood. Here they come closer to Radin's interpretation of primitive society in which the ritual drama serves to structure adolescent resolutions of developmental conflicts by helping the initiate to appreciate the shared aspects of his experiences without depriving him of a sense of individuality. Rather than repressing powerful infantile affects and conflicts, primitive ritual and mythology serve to raise these conflicts to awareness and facilitate their sublimation in adult roles.

Erikson's discussion of the psychosocial moratorium parallels Radin's conception of the ritual drama in that both agencies serve to preserve continuity in the adolescent's sense of self by helping him to reconcile early identities with later roles. Problems arise when a society fails to provide a secure interval in which such reconciliation can take place. Under these circumstances, infantile fixations remain the main determinant of adult character without the leaven provided by experimentation with various identity models. It is the lack of a satisfactory psychosocial moratorium that helps to account for the prevalence of symptoms like boredom and apathy. These complaints refer to the quality of adult experience and signify character disorders.

Most of the concepts used to describe contemporary American character reflect an awareness of this problem. David Riesman's hypothesis about the shift from "inner direction" to "other direction" captures this dimension. The other-directed man is characterized by his lack of any distinctive identifying qualities beyond sensitivity to the expectations of his fellows, coupled with the ability to appear to conform to these expectations whatever they may be. Erich Fromm defines this last capacity neatly in his label for modern man, the "marketing" orientation. Both of these social psychiatric theorists view this personality orientation as a response to the rapidity of social change in America. They view the "flexibility" that characterizes this personality type as constituting a kind of adaptation to the overwhelmingly rapid change in models and values.

Neither of them is satisfied with the quality of the adaptation so that Fromm hopes for greater "productiveness" and Riesman looks forward to increased "autonomy." Yet neither is able to specify in any concrete forms the social conditions under which these favorable outcomes are likely to occur.

Here, Erikson makes a real contribution. He recognizes the same condition described above as "other direction" and "marketing" orientation in his conception of "identity diffusion" which refers to a condition normal during adolescence but pathological when it persists beyond this phase. In a sense, Fromm's and Riesman's adapted types really perpetuate "diffusion" since the sources of identity lie exclusively outside the individual. People learn to manipulate the images they present to others so as to guarantee that they will be accepted, but they fail to establish any distinctive personal identity. Their childhood is never linked to their adult roles because they did not have any proper opportunity to experiment with alternative self conceptions during adolescence. Sullivan shows how anxiety serves to blot out perception of alternative life styles by forcing dissociation of personal impulses and selective inattention to threatening social options and models. The person coming to adolescence without satisfactory resolution of the juvenile phase lacks the psychic flexibility to experiment with non-stereotyped identities. Real intimacy is inaccessible and insecurity suffuses interpersonal relations.

On the sociological side, contemporary adolescents have their identity problems aggravated by the lack of stable models and satisfactory moratoria. Most American adolescents are exposed to several conflicting directives. Adults are frequently so insecure in their own identities that they prove incapable of giving genuine guidance. Parents create complications by projecting their own unrealized aspirations onto their children with sad results. Identity models provided by the mass media which cut across class and sub-cultural lines are usually so abstract and fantastic as to become little more than outlets for wish fulfillment rather than workable alternatives.

Our earlier analysis of American community studies provides an interpretation of the setting in which these identity conflicts arise. The segregated ethnic ghettos of the twenties usually sheltered the older immigrants from excessive contact with American patterns but they had no way of insulating their children to the same extent. The second-generation revolt against the ghetto led to the establishment of areas of second and third settlement along Americanized lines. But the grandchildren of immigrants find little to revolt against or to stand for and consequently suffer from "diffusion" rather than the "marginality" that plagued their parents. Industrialization in Muncie destroyed the old craft identities by removing their occupational moorings. In Newburyport, the breakdown of traditional status systems in the town and the factories challenged old identities and patterns of relationships between groups, without substituting any clear-cut alternatives. This early "phase" in American community development was dominated by the decline of traditional class, ethnic, and community identities to which the older generation clung compulsively, leaving the younger generation committed only to vague images of worldly success.

Part II takes as one of its central themes the identity problems of members of various American communities from the twenties through the present period. Severely restricted options open to young adults in the slum were illustrated by examining the life styles of two different kinds of corner-boy gang leaders, Doc and Chick. The alternatives confronting Bohemians in Greenwich Village during the twenties were then explored. Southern regional identities are taken up next with special reference to the artificial polarities imposed by the caste system. Then the identity transformations occasioned by military service during World War II were described.

The contemporary suburb exemplifies the social context in

which "other direction" and the "marketing" orientation are most likely to find institutional reinforcement. Park Forest offers the clearest instance of these character types, but Crestwood Heights and Exurbia exhibit their own version of the problems created by "adaptations" of this kind. Adult exurbanites are torn between their secret dreams and the realities of their lives to such an extreme degree that they find it hard to accept their jobs or their family lives as meaningful or worthwhile. Crestwood Heights males are more satisfied with their lot but the females appear at least as restless as their exurban counterparts.

Psychoanalytic theories provide the community sociologist with two kinds of perspective toward his subject matter. They permit exploration of patterns of medical psychopathology in a community, on the one hand, and they facilitate the study of the emotional consequences of "normal" psychic functioning, on the other. Psychoanalytic developmental schemes help to organize data about the conflicts and accomplishments accompanying human growth, and provide a way of looking at the life cycle which complements the anthropological perspective taken up in Chapter 10. The central position of adolescence underscored by recent psychoanalytic theorists supplements Radin's conception of the part played by puberty rites in a primitive society. Both frames of reference provide clues to subtle structural problems in modern American community life and accordingly broaden our conception of contemporary "disorganization."

Modern man's quest for identity is complicated by all of the changes accompanying urbanization, industrialization, and bureaucratization. Psychoanalytic therapy offers great hope to those who can meet its stringent financial and emotional requirements, but the "reality" problems that arise in communities, offering no moratoria to their adolescents and no opportunities for individuated growth to their adults, can hardly be solved in their entirety in the psychiatrist's office. Perhaps the

[273]

"other-directed" adaptation is a way-station enroute to a more complicated personality type, whether it be called "autonomous" as does Riesman or "productive" as does Fromm. These "healthy" types share the quality of being able to shape their social roles creatively rather than being dominated by them. This obviously requires changes in social opportunity as well as in the people to whom the opportunities are offered. Whether "productive" communities precede "productive" personality types or vice versa seems far less important than the larger problem of learning to create the conditions for encouraging both.

SOCIOLOGICAL PERSPECTIVES ON THE MODERN COMMUNITY

EACH of the chapters in Part II showed how the processes of urbanization, industrialization, and bureaucratization shaped the social structure of a different type of community. The fact that these separate studies, undertaken independently at different times and places, revealed underlying similarities at the level of community processes tends to confirm the generalized significance of the shaping processes that form the basis of this theory. These processes are still transforming American communities. This concluding chapter will outline their general effects focused around the social structure of suburbia. For purposes of contrast, a recent study of a rural village is introduced. Though the material considered hardly lends itself to summation as a series of logically related propositions, a coherent perspective on past and present community trends is presented.

Improved theorizing is clearly dependent on further empirical investigations; more studies like *Crestwood Heights* are indispensable. There is every reason to suppose that novel structural patterns will be found. The task of the theorist is to identify points of continuity between such emerging patterns and those which preceded them. We will be exploring the character of urbanization, for example, in the fifties as distinct from but closely related to Park's exploration of related processes in the twenties. The two are indissolubly linked by the fact that the bulk of the adult generation in the fifties was raised in the cities of the previous decades. Within a given time span, there will be this minimal continuity of personnel. Thus present-day suburbanites may well have been reared in the ghettos of the twenties and their behavior with respect to their children as well as to other areas of institutional participation is certainly shaped to some degree by their reactions to the sub-communities in which they spent their formative years.

For this reason, no community can be studied in static cross-section, not even a new suburb which has no apparent "past." At a minimum it has the past community experiences of its adult members. Whether homogeneous or heterogeneous, these influence the present community. Clearly, in the present stage of urbanization, the likelihood has greatly increased that people with diverse past community experiences will congregate in an area of common residence. As the ethnic ghettos disappear, their former inhabitants now settle in neighborhoods according to their social class. Distinctively ethnic institutions disintegrate or lose their significance in the experience of the assimilated second and third generation. There is some evidence that present-day areas of third settlement display only a minimum of distinctive ethnic patterns, with the life style of the community being mainly class determined.

One interesting effect of the blurring of ethnic lines is the loss of any sense of distinctiveness on the part of adolescents. Raised in a community where religious or national backgrounds have been forgotten or deliberately ignored, they lack any sense that their way of life is different from the standard American package, and indeed, for the most part, it probably is not. As in Crestwood Heights, Jews are virtually indistinguishable from Protestants in any terms beyond details of religious belief and ritual. Similarly, most third-generation Italian youngsters act pretty much like their third-generation Irish and Jewish counterparts. One result of this is that the adolescent sense of participation in any community other than American mass society dissipates. Ethnic identity is no longer a matter of concern to them. They acknowledge it far more easily than many second-generation ethnics who struggled desperately with their cultural past, perhaps because doing so entails no commitments that might cause them to be regarded as "outsiders."

College teachers often bemoan the lack of any real appreciation of the experiences of recent generations of Americans shown by their students. Yet this is an inevitable consequence of the rapidity of social change and the standardization of mass society. How are these youngsters to find out that their fathers and mothers were raised quite differently? Perhaps some responses from my own students at Brandeis might be mentioned, since they are, for the most part, children of suburbs. When this question was raised in a class of thirty, most of them from upwardly mobile Jewish backgrounds, only one had ever seriously entertained the idea that his parents might have experienced different community patterns from those with which he was familiar. This enterprising student had gone so far as to mention the matter to his father on one occasion; but the father, who had come from an immigrant slum, not only refused to admit the differences, he refused to discuss them. Perhaps he was being more honest than most, since the conventional gambit is to present a sentimentalized version which leaves the child more confused than before. This tendency to see recent history in terms of fashionable middle-class imagery includes a set of stereotyped images about earlier decades. Two forms of stereotyped nostalgia, idealization of the "lost" generation of the twenties and idealization (or its mirror-image, denigration) of the "radical" generation of the thirties, commonly appear among students and even among some adults who ought to know better.

These same students who find it hard to conceive that the recent past could have been significantly different from their version of the present also have trouble conceding that their personal experiences could differ significantly from those of their classmates. Apparently the sheer fact of idiosyncratic experience has become inaccessible and even threatening.

[277]

These students in their private reveries anxiously exaggerate the "uniqueness" of their personal experiences because they have few social opportunities for affirming, reinterpreting, comparing, or sharing them. Their accustomed system of age-sex categorizations leaves little room for comprehending individuality. This might be a reduction to an absurdity of one conception of equality: not only are people equal, they are exactly alike. The roots of this leveling process go deeper even than the oft-noted "conformism" of the college generation. It is part of the psychic set from which conformist attitudes arise. When the past is, as it were, reduced to the present and all experience in the present leveled to that of a standardized middle-class American adolescent, the perception of alternatives prerequisite to nonconformity has been eradicated. Like the father of the student mentioned earlier, many adults unconsciously collude in perpetuating this world-view and there is no doubt that the mass media contribute greatly to reinforcing it. Even when television, for example, opens "windows on the world," the men in charge "give the public what it wants" by making certain nothing unacceptable to the cliché system of the middle-class comes into view.

Park, in his interpretations of city life in the twenties and thirties, noticed this "blindness" toward sub-cultural differences as had Simmel before him. They both regarded it as part of that sense-deadening "sophistication" necessary to keep from being overwhelmed by the multitudinous stimuli of city life. It was this which permitted people with utterly different and even conflicting moral codes to live side by side. When they met, their relations were kept highly impersonal so that neither need confront the values by which the other lived or feel outraged by them. The true urban sophisticate possessed an impersonal "presented self" that allowed him when necessary to interact with almost anyone while avoiding deeper relationships that might disturb either party.

These students would seem to represent an advanced stage

of urban sophistication. Their blindness to socio-cultural differences has involved them in a corollary blindness to historical processes and to unconventional individual differences. Since the twenties, the vocabulary through which people can converse together on an impersonal basis has been greatly enlarged and its diffusion to all elements of the population highly accelerated. We are all so accustomed to "interacting segmentally" that we find it hard to do anything else. Yet the sophistication of our students is different from that predicted by Simmel and Park, in that they expected an awareness and even acceptance of various moral worlds while the contemporary sophisticates ignore or deny these.

One kind of summation of the theory of communities presented in the earlier sections is the tracing of community experiences of America over the past thirty years in terms of the studies and concepts developed therein. The cities of the twenties were extremely heterogeneous entities composed of more or less culturally distinct, spatially segregated sub-communities. One's perspective on the city was that of one's sub-community and even the mobile were able to maintain some contact with their original sub-culture by congregating in the area of second settlement. But the commitment to sub-cultural values was already being weakened by the prospect of "Americanization." Even in towns as small as Newburyport, ethnic participation in and absorption into the American social system was beginning to weaken ethnic institutions and identities. The drift was toward standardization in the direction indicated by mass media stereotypes of middle-class America. Park's concern with the need for impersonal controls to prevent disruptive deviation becomes less cogent, until by the fifties these impersonal controls threaten to destroy the very diversity that once made city life attractive.

A step in the diminution of urban diversity is the disappearance of Bohemias. Caroline Ware provided a glimpse of the moment in the life of Greenwich Village when its transformation from a genuine to a false Bohemia was imminent. Here we can identify the mechanisms through which the very amorphousness and quest for marketable novelty dominating American life rendered the consolidation of distinctive Bohemian patterns difficult if not impossible. The experimental family was taken over along with experimental art, though its meaning in the suburban context was quite different from that in Bohemia.

Park's students familiarized us with slum life in the twenties. Whyte helped to correct certain impressions about "disorganization" by exposing the distinctive social organization arising in ethnic lower-class neighborhoods. There are still a good many Americans living in neighborhoods properly classified, though often not by their residents, as "slums." Research in such sub-communities is often dominated by middle-class images, and it is here that the structural models of studies in earlier periods can themselves become blinders. Whyte has shown this to be true of Park's conception of slum structure and he provided a detailed picture of the institutional realities of slum life along with their human meanings to various categories of slum dwellers. Whyte shows that slum life styles and life plans are shaped by forces related to those found in Muncie and Newburyport.

Industrialization, spreading around the turn of the century, revolutionized the work process by requiring degrees of specialization that eliminated most vestiges of coherence or craft satisfactions from the work process. In addition, as Lynd has shown, it ripped apart the fabric of communal life by putting a premium on the superior strength and speed of the youngsters while devaluing the laboriously acquired craft skills of their parents. Meaning and purpose had to be sought outside the factory, but the more personal sources—religion and the family—were also in the throes of change. The teachings of the church became more and more remote from the activities of life while new economic necessities and opportunities overthrew established family patterns. Lynd indicates that dedication to an ever-rising "standard of living" was the only way to justify industrial work roles, though the commitment of the older generation was always tinged with nostalgia for a craft way of life. The younger generation, those now between forty and fifty, knew this other pre-industrial round of life only tangentially if at all. They were prepared to renounce "intrinsic" work satisfactions and, for that matter, practically any other kind of intrinsic satisfactions, so long as they could gather the emblems of economic success.

The people of Newburyport also climbed onto the mobility ladder. Their transition from a traditional, old-family-dominated New England town to an absentee-owned class-divided community brought with it the crucial experiences of the Middletowners, but heightened by simultaneous disintegration of a traditional system of social organization that had satisfied important needs. As Warner shows, prior to bureaucratization there was a community consciousness in Newburyport that allowed the old families to provide leadership that symbolized the aspirations of the whole community in a fashion rarely approximated in American life.

The American sense of remoteness from power centers was greatly accentuated by experiences during the depression when it became all too clear that national and international forces well beyond the purview of the local community determined life chances to an important degree. In Middletown, the people were forced to turn to the federal government for help and this established a pattern that has since persisted. The Lynds' most striking observation about the effects of the depression— that it actually heightened affiliation with the success formula —is certainly borne out by the present behavior of men who were young adults during that period. One cannot but wonder if some of the underlying anxiety and refusal to recognize the past is not a result of the traumatic circumstances suffered by persons committed to mobility during the depression. Certainly it must have contributed to their foreshortened time perspective.

World War II brought another set of difficulties. Most people were mobilized either in the army or in civilian war jobs so that pursuit of careers was seriously interrupted. At the same time, the Army itself provided experience with an authoritarian bureaucratic caste system that aroused almost as much resentment and anxiety as the war hazards themselves. Commitment to the military program, which could have been the occasion for arousing deep national sentiments, seems to have evoked mainly individualized responses. The war period became but another in the succession of upheavals and interruptions in the all-important upward climb.

In this historical context, neglect of the past by present-day adolescents and repudiation of the past by their parents becomes intelligible. More so than ever before, energies have been thrust into creating a worldly paradise based on material acquisition. This version of the American Dream is embodied with greatest clarity in the prosperous suburb and can therefore be studied most conveniently in that setting. *Throughout the following discussion, the point of reference will always be the suburb unless a larger context is specifically indicated.* Many of the observations about suburbia can be extended to other sub-communities with only slight modifications. As the suburban style increases its hold on the dreams and the lives of Americans, the usefulness of this sub-community as a social laboratory increases correspondingly.

The social structure of the prosperous suburb is strangely paradoxical. On the one hand it arranges matters so that the daily life of the individual, no matter what his age or sex, is divided into many compelling tasks that leave little or no time for freely chosen activity. Like modern industrial employment, which it fundamentally resembles, the suburb is frantically devoted to the rhythm of keeping busy. Even the playtime of the children is routinized and many families find that the separate schedules of the various members leave no time for intimate moments with one another. On the other hand, while people are so desperately busy, they do not know or have forgotten how to perform some of the most elemental human tasks. The most glaring evidence for this is the genuine crisis in the American suburb over child rearing. Anxious mothers, uncertain how to raise their children, turn to scientific experts to find out if their children are normal or, even more important, what the standard of normality might be.

We face a curious and probably unprecedented situation here: a society of material comfort and apparent security in which the most fundamental of human relationships—that between mother and child—has become at the very least problematical. No one is surprised to discover that businessmen treat each other in impersonal and manipulative terms; but surely it should be cause for some dismay to find it habitual, as the authors of *Crestwood Heights* report, that mothers regard suburban children as "cases" the moment they lag behind the highly formalized routine accomplishments of their peers or, still worse, shows signs of distinctive individuality. The paradox between busyness and helplessness, between outer bustle and inner chaos, may now become easier to explain: "Keeping on the go" is the prime way for the suburbanite to avoid facing the vacuum in which he lives. Hence, the peculiarly painful fact that in suburbanite society, for all that it is so conspicuously child-centered and for all that parents habitually make sacrifices in order to get "the best things" for their children, it is an unusual mother who really knows her own child.

For that matter, no one in the suburb really has to know anyone else as long as appearances are kept up. Housewives, taught to desire careers, are trapped in the home. Husbands, trapped in careers which drain their best energies, must look forward to a fate that has become as dreaded as death—that of retirement and free time. Looking ahead to their own prospective life cycles, the children soon learn to submerge the specter of a life that lacks rooted values and creative meanings by throwing themselves into the struggle for status. All of the vital roles wherein the human drama used to be played out—mother-son, father-daughter, worker-player, adult-child, male-female—now tend to be leveled. Their specific contents which had previously made them into channels for realizing a particular set of human possibilities have been bartered for an ephemeral and empty sense of status. Not to perform these roles is to lose one's place but, sadly enough, performing them can never give one a place.

Now it is true that all societies use status, the systematic allocation of prestige and esteem, as a way of motivating people to fill social roles. Those performing valued activities in a competent manner receive the awards of respect and approval. This presumes, however, that there *are* some stable roles which receive acclaim. Crestwood Heights, however, is distinguished by the absence of this kind of consensus. For not only does the suburbanite have to keep up with the Joneses; he also has to spend a tremendous amount of time and energy trying to find out which status models the Joneses are themselves currently keeping up with. Since no other activities are finally valued in and of themselves apart from this quest for status, it soon becomes obsessive. *Status becomes an autonomous motive and mode of life.* Human relationships are valued only as sources of status when status ceases to be the reward for having successfully become a valued kind of human being. We have reached here the result of that process within bourgeois society which, call it alienation with Marx or *anomie* with Durkheim, transforms the human being into an object—and this strikingly enough, at the very moment when human beings in the suburb believe they have triumphantly won the battle to gain control over objects.

From a social psychiatric perspective, the suburb presents many interesting problems. It appears that the character types and mechanisms considered most prominent in contemporary

[284]

American life are often actually best exemplified in suburbs. There, if anywhere, would be found Erich Fromm's "marketing orientation," David Riesman's "other-direction," Harry Stack Sullivan's "exaggerated security operations," Erik Erikson's "identity diffusion," and even possibly the thoroughly unsavory "authoritarian personality." There are broad convergences among these various concepts deriving at least in part from the fact that the personality configuration described by all of them emerges from what can conveniently be called "mass" society. And the suburb is a better place in which to study this human type than most other city neighborhoods, even though all of them contain some form of the species.

Chapter 9 summarizes the descriptions of personal relationships in Crestwood Heights. There is massive covert competition between children in the school, women in their clubs, and men at their businesses. Everyone is trained to deprecate his competitors without appearing to do so. As they rise through the complex status structures that make up their worlds, careful manipulation of their own personalities remains the one indispensable stock-in-trade.

All social psychiatric descriptions of personality types finding their natural habitat in suburbia emphasize the central role played by anxiety. This anxiety is rarely recognized as such but it need not be recognized in order to govern effectively the suburbanite's self-marketing operations. The children become wary of its sting when they discover that they are part of the status equipage of their parents and must comport themselves accordingly. They are loved for what they do rather than what they are, and this becomes especially bewildering and devastating because their parents constantly insist that the opposite is the case. Small wonder that the child begins to see himself in depersonalized, almost "schizoid" terms.

It should be stressed, however, that a term such as "schizoid" loses its specifically psychopathological implications when we deal with a community in which this response could hardly be

[285]

avoided. The anxieties that cue such responses *must* remain well below the threshold of awareness; otherwise, they would leave the individual in an unbearably agitated state. Children and adults alike develop standardized patterns for restoring some kind of inner balance when their security is threatened; however, it is clear that the price paid for this restored equilibrium is an impoverished emotional life. Because they cannot face their own anxiety and its origin in their life routines, they are unable to visualize other solutions.

The situation quickly becomes self-sealing. Adults in the community, committed to precariously maintained status postures, must continually reiterate to each other and to their children proclamations of "success and happiness." To do this well, they have to delude both themselves and their audiences. The child must appear "happy" so that his parents' success as child rearers will be confirmed. However, the real feelings, insecurities, and dissatisfactions of the parents are conveyed in one fashion or another, perhaps only non-verbally, to the children. They are even likely to sense the complex network of exploitative emotional entanglements that undercut the tranquilized surfaces of their family life. But there is nothing they can do with this recognition except suppress it, since everything else would involve repudiating the only security-giving group they have. They cannot even find out if other families are like theirs in this respect because the accessible family groups, those of friends and relatives, will always seem to be hidden behind the stereotyped masks. To decide, as some children seem to, that their own family is inadequate or even hypocritical leaves them exposed to deeper wounds. Otherwise, these doubts and the underlying realities are avoided and the child inducted into the ways of responsible adulthood.

At this point, we can clearly see the most grievous human loss that life in the suburb entails. Dedication to a status-dominated life style forces individuals into a rigid mold from within which they can see only limited aspects of human reality.

Other people become threats or objects to be used. Emotional growth stops at Sullivan's "juvenile" phase. The identity struggles of adolescence are resolved through stereotypes that simplify reality rather than through fresh perceptions that provide a basis for expanding contact with personal and interpersonal realities.

Yet it would be a mistake to assume that the typical suburbanite's anxious preoccupation with status as well as his anxious avoidance of genuine experience insulate him completely from life's conflicts. He may be able to postpone the conscious facing of some conflicts but he cannot permanently escape from their consequences. Women usually suffer from social contradictions more than men because they are confronted squarely and inescapably with at least one role conflict. Whether they choose housekeeping, careers, or a combination of both, they must sacrifice something important. Their busy schedules prevent them from finding too much time on their hands in which to reflect on their situation but those who do can always turn for help to one or another of the ubiquitous psychological experts who play an increasingly important role in Crestwood Heights. These experts begin to look like modern shamans exorcising the tribal ghosts so that the routines of living can be kept up without interference. But these shamans cannot touch the underlying causes of the anxiety they temporarily allay.

One suspects that the career-driven males occasionally feel doubtful about the whole enterprise. The barrenness of their lives cannot be completely concealed so that as they approach the "male menopause" this emptiness may even penetrate their atrophied sensibilities. As the age of obsolescence nears, the old compulsive preoccupations lose their power to divert attention from the realities of their existence. By then it is usually too late to do much about it.

It is important, however, to recognize that the "human material"—just because it is human—does not completely adapt to the pressures. Continual distractions provided by busy schedules and supplemented by heavy doses of mass entertainment may turn attention from the real problems of life, but the half-dreaded, half-desired moments of unavoidable reverie cannot be entirely suppressed. These realities can break in symptomatically as with psychosomatic conditions or they can manifest themselves interpersonally in terms of "problem" children or "problem" marriages. Even more devastating, at least for the male, is the much-dreaded work "block" which can threaten careers. And there are more conventional disorders—people can't sleep or sleep too much; they can't eat or eat too much. In these seemingly minor matters, the malaise takes serious toll.

When clearcut psychopathological or psychosomatic symptoms appear, they do open up, at least for some people, the alternative of psychotherapy. This can provide the disturbed individual with a new identity: he becomes a "patient." In this role, he is allowed to express deeper human feelings, including his loves or hates directed at parents and parent-figures, his multitudinous ambivalences, his forgotten fears and forgotten dreams, mourning and much else that has little opportunity for free play in the constricted contexts of ordinary suburban life. Some contact is established with the emotional profusion of childhood without running the risk that this will endanger one's foothold in the world of careers and career-marriages. New possibilities are indeed opened up but there is absolutely no way of ensuring that they can be transferred into the patient's everyday life. If he chooses to remain in the suburb the life order is such as to militate against this carry-over; and so he is forced to prolong his therapy since that has become one of the few relationships in which his "cure" can be manifested. The patient's life commitments undo many beneficial effects of the therapeutic relationship and this paradoxically renders the patient even more dependent on it. Perhaps this helps to explain why so many analyses drag on endlessly.

There is no reason to assume that the problems of identity diffusion and foreclosure described above stop at the boundary of the suburb. Few people reared in modern America manage to escape the pressures toward espousal of stereotyped self identities as a means of gaining a temporary resting place in the unending status struggles, even though the price may well be emotional and intellectual stagnation. The role conflicts of suburbanites in which elemental human identities and problems are distorted by the quest for status are by no means their exclusive prerogative. Establishing satisfactory masculine identities at each phase of the life cycle is as precarious among workers as among managers, while the wives of both groups struggle with the problems of femininity. Unfortunately, it is always easier to settle for the stereotyped juvenile patterns promulgated by the mass media, encouraged by an inner monitor, the self-system, and reinforced by the demands of modern social life. Still, growth does take place. People do occasionally find their way through the banalities of mass society and it is important to study with great care the conditions under which this occurs. Before individual progress can be made, identity must first be seen as problematic and the growth anxieties as real.

A recent study of a rural up-state New York town having a population of 1,700 shows interesting differences from and similarities to suburbia. Arthur Vidich and Joseph Bensman, in *Small Town in Mass Society*, present a carefully drawn picture of the way in which urbanization, industrialization, and bureaucratization have shaped the social system of a rural village. Their findings are important for this theory because they show how the same organizing and disorganizing forces at work in suburbia are changing a community which has long been regarded as being the last outpost of resistance to these forces—the small town.

Small towners still cling to the traditional image of their community as a stronghold of opposition to urbanism. Indeed, this conviction forms the central theme in their self-image. Springdale takes great pride in its "neighborliness, equality,

[289]

grass roots democracy, and rugged independence." This favorable self-conception is juxtaposed against a counter-image of the big city as a place steeped in corruption and devoid of human values. The townspeople's imagery is reinforced by those portions of the output of the mass media to which they pay attention. And no one dares challenge the small-town myth without running the risk of exclusion from local affairs.

Research showed that this public image had very little relationship to social reality. Vidich and Bensman found that the town contained a set of socio-economic classes with clearly differentiated life styles and life chances; judgments of personal worth rendered by Springdalers were based almost entirely on economic success; there was no indigenous local cultural life worth mentioning; political affairs were dominated by a single top leader who used three lieutenants to control all of the elected political offices and who made all important policy decisions; and finally, even this group of "top leaders" actually served as mediators between Springdale on the one hand, and the higher echelons of political authority at the state and national levels, on the other.

This disparity between public imagery and social realities put the field workers in an odd position. The familiar and indispensable sociological injunction to explore the meanings that social structures hold for the people whose behavior they pattern led them up a blind alley. Since the essential structural features of the community, through which its affairs are linked to the larger society, are not recognized by the participants, these structures cannot possibly hold any meaning for them. The local class system, the local political machines, the dominance of mass culture were all not only unrecognized, but when attention was called to them, their existence was vigorously denied by the very people involved. This appears to be a special kind of "latent" social structure. The elements are not only unrecognized and unintended, but the mere suggestion that they exist arouses intense antagonism.

Refusing to be put off by these denials, the sociologists continued their observations. They accumulated detailed evidence of the ways in which "metropolitan dominance" manifested itself in Springdale. Every aspect of local life depended upon adjustments to mass society. But these adjustments were rarely recognized as such by the people making them. Only a few marginal respondents could identify the true state of affairs. The sociologists were compelled by their findings to focus on this "latent" adaptive social system as one of the main problems of their study.

Vidich and Bensman describe two dimensions of social control in Springdale. The first dimension is the maintenance of belief in the small-town ideology, while the second is the maintenance of a social system that adapts local affairs to the pressures of mass society. Since the local ideology categorically denies the existence of any adaptive system or of the need to adapt in the first place, it would appear that these two dimensions are working at cross-purposes. The authors emphasize the way in which this contradiction between ideology and reality serves as the focus of both organizing and disorganizing forces in Springdale. Their problem is: "How can social integration and personal integration be preserved despite the potential disorganizing effects of the contradiction?"

Social integration is analyzed as a problem in leadership. The four top leaders assume primary responsibility for keeping Springdale in touch with state and national power centers. Working behind the scenes, they make certain that the village and town boards reach "proper" decisions. They have created a political machine in which other local people, including the elected officials in town, serve as secondary leaders without real policymaking powers. The top leaders dominate the school board; though the school principal, who is an outside professional, manages to achieve some small degree of independence. Churches organize much of the social life of the community and the ministers also defer to the "invisible government."

[291]

Integration at the level of community action is achieved by this autocratic power structure through which local affairs are regulated and the demands of mass society met.

But the existence of this "invisible government" could constitute a threat to social integration if the local citizenry were ever to recognize its workings and realize that these make a mockery of their assumptions about "grass roots democracy." Such recognition is prevented and personal integration brought into line with social integration by the habit of "particularization"—a concept used by the authors to characterize the way in which Springdalers perceive social processes and events. It consists of seeing events and processes in isolation from one another so that generalizations cannot be drawn. It is actually a partial paralysis of the faculty for abstract thought. Springdalers are able to encounter any number of specific facts attesting to the existence of the political machine, class differentials, or the town's dependence on mass society, without ever recognizing that these encounters indicate the existence of important regularities. Idealized conceptions are never tested against realities because the faculty for perceiving these social facts has been extinguished.

Even this perceptual habit would probably not be sufficient to inhibit awareness of social realities were it not reinforced by a round of life in which little opportunity appears for introspection. Signs of growing self-awareness are considered danger signals indicating "abnormality." The authors call this tendency "externalization" and they analyze a number of institutional mechanisms that sustain it. The most important is, of course, the high value placed on "hard work." There is also constant pressure toward activity of all kinds. Church activities and organized political life involve heavy time commitments. Springdalers keep every bit as busy as the schedule-bound suburbanites in Crestwood Heights.

Vidich and Bensman offer a detailed interpretation of the control mechanisms on the institutional and the social psychological levels. From the standpoint of this general theory, the details are less interesting than their overall finding—that is, a community can respond to the pressures of urbanization, industrialization, and bureaucratization by forcing its members to adjust to the institutional necessities imposed by those pressures, while at the same time forcing them to deny the existence of the pressures and their structured adjustments to them.

Springdale provides a case study of an extreme response to mass society that highlights some interesting features appearing to a lesser degree in other communities. The myth of local autonomy is not peculiar to the small town though it is not usually as central an aspect of community ideology. The distinctive feature is the organization of community affairs around this myth to a point where the defense of this illusion entails serious impairment of symbolic functioning. It is the main source of self-justification for the community as a whole, and for its individual members. Challenges to its validity threaten collective and personal security at the deepest level. It is less painful to disregard challenging facts than to confront their implications. And the illusion of complete local autonomy, like the illusion of complete personal autonomy, is sustained at the price of subordination to the demands of mass society.

Awareness of mass society takes different forms in each of the suburbs studied. The exurbanites are at the opposite pole from Springdale in one respect—they are painfully aware of their subordination to technical roles in mass society. Their movement to the exurbs involved a response to the constrictions imposed by these roles. They sought a synthesis of rural and urban values but found themselves plunged even more deeply into the "rat race" because of the heavy expenses required to maintain their rural facilities. And they remain too sophisticated to avoid recognizing their plight by such mechanisms as particularization, nor can they lose themselves in external activities. No taboo on introspection could be enforced

[293]

in a community so heavily committed to psychoanalysis. Their adjustment takes the form of institutionalized "cynicism" which contrasts sharply with the institutionalized "naïveté" of the small town. This cynical posture allows them to feel that they are distinctive because they know the real truth about modern society and therefore are superior to the "suckers" who have not seen through the disguise. Despite their "cynical" masks, many exurbanites cherish a secret dream of turning to "creative" jobs. But their mounting indebtedness keeps pace with their rising salaries to eliminate this as a realistic possibility.

Park Foresters appear to be quite aware of the workings of mass society. They are committed to a search for secure niches within the large-scale organizations that employ them and seem willing to pay any price, including conformity, for this refuge. The social system of Park Forest with its enforced sociability and its marginal differentiation hardly causes concern to most persons participating in it. William H. Whyte's book *The Organization Man* probably held few surprises for the well-trained organization men who read it; indeed, some of them must have wondered what all the shouting was about. And, as Whyte shows, there are distinct satisfactions in "transient" community life, not the least being the close ties and sense of shared experience which they can never recapture when they move to more well-to-do but more impersonal neighborhoods. Park Forest as a typical phase in the career cycle of executives provides necessary opportunities for the acquisition of values, attitudes, and social skills that facilitate role performances in later stages of their careers.

Crestwood Heights is one of the community systems that successful executives enter. Its residents are independent professionals, businessmen, and upper-level executives, all of whom have "made the grade." The institutional system differs from Park Forest by providing more occasions for privacy but there is equal emphasis on "keeping up appearances." A new element now enters. The value system emphasizing "maturity" and "self-realization" which plays a rather small part in the educational and family lives of the transients is more prominent in Crestwood Heights. Conflicts arise between these "maturity" values and the dominant "success" values. There is a tendency to resolve these tensions along the same lines that the contradictions between ideals and realities are resolved in Springdale. Children in this suburb learn to strive for competitive success while proclaiming their interest in maturity. The contradiction creates some overt conflict as the "progressive" ideals are upheld by two sub-groups—the women and the experts. But these sub-groups are themselves deeply ambivalent since they have not rejected material success as a life goal. There is widespread "inner" conflict as well as conflict between representatives of the competing life programs.

All three suburban communities display a degree of sophistication about the workings of mass society that distinguishes the three of them from the precarious "naïveté" that prevails in Springdale. Whether they accept domination by mass society, as do the conformists in Park Forest, or try to struggle against it like many exurbanites, their activities are based upon some insight into the effects that it exerts on their own affairs. By proclaiming its total victory over urban trends, the small town deprives itself of any opportunity for reasonably appraising its own position in the modern world.

Taking Springdale as a case in point, it is hard to see how the small towner could gain much by being able to appraise his situation rationally. Given the conditions of rural surrender, perhaps the retention of a mythology that allows a minimum of collective self-esteem and justification is the only real alternative to social disintegration. People need to believe in the value of the communities in which they live, the goals that they seek, and the satisfactions they receive. None of this would be possible if the true state of affairs was exposed; and it is ironic that exposure would not necessarily render the facts amenable to change. Perhaps local political affairs could be

[295]

made more democratic, but the power in the hands of the invisible government rests on a solid socio-economic basis and would not be taken away easily. Certainly "urban dominance" would not be diminished if its existence were known—any more than their intimate familiarity with the "rat race" helps the exurbanites to extract themselves from it.

There is a terrifying finality about the facts of community organization when they are approached from the standpoint of an objective observer. The processes of urbanization, industrialization, and bureaucratization at work in mass society appear to shape the destinies of communities and individuals along irrevocable lines. We watch the doomed craftsmen of Muncie, the doomed old families of Newburyport, and the doomed first-generation immigrants go their respective ways toward oblivion. Struggle seems futile because the power and impersonality of the forces at work makes them impregnable. Community ties become negotiable commodities as the mobile middle-class climbs its way upward. Their highest value is "keeping up appearances" and their highest goal the acquisition of fashionable commodities. Working-class acquiescence in this life program has not lagged behind. And all the communities in mass society look like so many minor variations on a single major theme. The students of social character reinforce this impression. David Riesman shows how the "other-directed" man fits into a changing social system with less strain than his "inner-directed" predecessor. Erich Fromm somewhat more critically documents the rise of "marketing" personality types in a market society. Large-scale organization creates small-scale personalities, and meaningless jobs are filled by eager jobholders.

Yet this picture is at least partly an artifact of the distance between the observer and the community. Anyone reading a large number of community studies can hardly help being impressed by the range of levels of abstraction in the data presented. The battery of techniques used by sociologists studying communities can be seen as arrayed on a continuum ranging from those providing highly impersonal data to those focusing directly on concrete individuals and groups. To illustrate, ecology and demography would fall near the impersonal pole while life-histories and the kind of participant observation employed by William Foote Whyte clearly focus on specific people and events. Park was concerned with gathering both kinds of data and so were the Lynds. Even so, the only people mentioned by pseudonym in either Middletown volume were the X and Y families and both are treated in a sufficiently abstract fashion to reveal little about the kind of people they were in either instance. Warner moves closer to describing individuals in his vignettes. It is not hard to see why families in Newburyport were offended by his revelations; he supplies enough detail so that most members of the appropriate class could easily picture themselves behaving in the manner he reports. William Foote Whyte furnishes us with so much information about both Doc and Chick that we can almost comprehend the personal motives that shaped their behavior. The more distant view of Doc and Chick might well have lumped them together as "slum gang leaders" without detecting the crucial differences so carefully explored by Whyte.

Along this same line, one wonders if there might be equally significant differences between the individuals and families of Crestwood Heights. Seeley, Sim, and Loosley, chose to pitch their investigation at a rather high level of abstraction which necessarily involves losing sight of individuals and particular groups. This was entirely proper as a research decision, but one wonders if it may not involve too easy acceptance of the assumptions about suburban standardization held by many members of the community as well as by the research team itself. If the sociological microscope had been tightened one turn more, might some chinks have been exposed? Perhaps the depressing sense of uniformity is at least in part an artifact of the research.

This does not vitiate Seeley, Sim, and Loosley's significant findings. Certainly opportunities for expression of individuality in suburbia, whether it be Crestwood Heights or Park Forest or the exurbs, are limited. Furthermore, character portraits or vignettes always raise difficult ethical problems, since the persons on whom they are based may be disturbed by them, or even worse, the real identities of the people involved may become public. Yet these are problems that must be struggled with since the absence of this kind of data leaves an important gap. Obtaining life histories from respondents who remain unwilling and unable to expose their life stories to objective scrutiny has similar hazards. Much time and patience will be required to encourage respondents who are themselves not inclined toward introspection or retrospection to tell the stories of their lives. Some satisfaction can be gained from realizing that this experience permits a mode of human interaction alien to mass society and therefore involves both observer and observed in an act transcending its limits.

This raises the obvious point that depersonalizing trends also affect the· scientist. Indeed, one flaw in the "march of history" approach outlined earlier is that it leads sociologists to ignore the private life space that people manage to secure for themselves. Even in the most conforming and standardized suburb, personal dramas still unfold. Anthropologists in primitive settings and psychoanalysts everywhere have their attention forced on these personal dramas, whereas sociologists are tempted to adopt impersonal role masks in their theorizing and in their research.

With all of Freud's pessimism about the destiny of civilization, he remained hopeful that rational insight could influence human affairs. Before this influence could be felt, individuals had to "work through" their early experiences and confront "repressed" conflicts. He devoted himself to studying these fundamental human conflicts, especially the Oedipus complex—with an eye toward diminishing their constrain-

ing influence on rational behavior. While civilization demanded ever-increasing instinctual renunciation, accompanied by increased guilt and anxiety, Freud strived to perfect
techniques that would permit civilized men consciously to
confront this anxiety and lead genuinely "personal" lives in
doing so. In this sense, he was as much an opponent of depersonalization in mass society as Marx or Weber.

Anthropological thinkers, especially Paul Radin, provide
comparative perspective to supplement psychoanalytic observations. Primitive community life can be viewed as a set
of arrangements for protecting the integrity of the individual
life cycle. Guarantees of a share of the necessities of life are
an integral part of the meaning of tribal membership. More
importantly, the tribe guarantees a share of the emotional
necessities by providing ritual celebrations of the phases of
life. The shared dramas of human life—the potentialities inherent in each phase of the life cycle, masculinity and femininity, the passions, ambivalence, man's yearning for omnipotence, his dependence on others and his quest for identity—
are the anchor points of primitive community life. Radin
shows how the ritual dramas and mythology create an opportunity for complex awareness of these dramas of socialization. Through the rituals, primitives are able to assimilate
their individual responses to these experiences without losing
their grasp of the range of alternative responses and without
blocking out their awareness of "anti-social" impulses.

Mass society, centered as it is around the unending quest
for social status, tends to level the phases of the life cycle.
Attention is forced on the identity conflicts connected with
social acceptance rather than those involved in individuated
growth. The immense joint effort that the psychoanalyst and
his patient must put forth to establish meaningful contact
with buried Oedipal strivings, for example, is a measure of
the distance that modern man has moved from the main-

springs of his life cycle. Yet these central human experiences are the basis for real differentiation and individuation.

It is important to remember that the community study is part of the personal experience of the investigator. Inevitably, his own community experiences provide a background for reading and research. It is necessary that this background be brought to awareness and its influence traced. In searching for community patterns affecting the lives of informants, the sociologist indirectly reflects upon his own life. Who can believe that Robert Lynd was not exploring the limitations and resources of his own boyhood in studying Middletown or that Whyte did not find in Doc and Chick a set of alternative "personae" closely related to his own life choices? To call attention to such "projective" meanings is by no means to derogate them. On the contrary, one suspects that in both instances their presence helps to account for the exceptional quality of the study.

So the sociological field worker has to combine self-insight, empathy, and objectivity to counteract the depersonalizing pressures which prevail in the intellectual community as well. Acquaintance with anthropology and psychoanalysis is helpful, and literature also offers useful models. Theodore Dreiser's novels deepen and reinforce Park's research; Sinclair Lewis' novels about life in the Midwest confirm the Lynds' observations; J. P. Marquand actually writes about Newburyport. The interplay of literary and sociological techniques for community exploration is quite intimate. Good novelists always show the implications of the events portrayed on the growth of their characters just as do good sociological life histories. City novelists have developed techniques for representing the multiplicity of urban experience which deserve attention. In USA, John Dos Passos interweaves news events, biographies of public figures, snatches of conversation, and reports on the lives of his characters to portray New York life around World War I in an impressive fashion. The imaginative act required for the reader to put these diverse materials together is similar to the imaginative act that any good sociologist performs when he synthesizes objective and subjective data. A developed literary imagination is a great help in doing, and no less in reporting, a community study.

Before a comprehensive theory of community comparable to the synthesis achieved by Park in the twenties can be put together, a far more complete picture of the levels of participation in and response to mass society must be obtained. This is a prerequisite for reformulation of Park's theory of natural areas. No amount of theorizing can identify the range of urban sub-communities in the modern city, nor can it describe their structures or their interrelations. Ecological and demographic data help to provide starting points for explorations into metropolitan social structures. Park's injunction to "get out and look" bears repeating. The single category "suburbia" contains at least three sub-types as exemplified by Park Forest, Crestwood Heights, and exurbia. There is good reason to suppose that many others will be found. Our cities obviously contain working-class suburbs, occupational communities, new kinds of business districts like the huge peripheral shopping centers, and even Bohemias.

Though we can no longer be certain that the "freedom of the city" is as meaningful as Park assumed, urban settings do provide opportunities for individual expression and association hardly possible in the small town. And theories about mass society should not be allowed to obscure the problems that this remaining diversity presents. It is likely that diffusion of standardized entertainment and information through the mass media exerts a homogenizing influence; however, this influence differs according to the community context in which it operates. Perhaps in some communities it affects public life styles, leaving intimate manners less affected. In others the reverse may hold true. Marginality is obviously still a common urban phenomenon, but its character has changed since the twenties. Fewer people are caught between two clearly defined cultural systems as were the immigrants and their children. Instead, the bulk of the population is exposed to many competing directives, none of which evoke very deep responses. But again the really interesting questions arise when the problem is posed in terms of specific sub-community contexts and the styles of marginality encountered in each of them.

Aside from its substantive inaccuracies, a monolithic conception of mass society stifles the research impulse. Why bother going to all the trouble involved in field work if the findings are known in advance? The main reason for carefully confining theoretical formulations to community contexts in this volume is to avoid such premature closure. Sociologists have to learn to let their theorizing and their observations range freely among various levels of abstraction without getting "fixated" at any single point. This is easy to recommend but hard to achieve. In this chapter, suburbia was discussed on two levels. The first was a general discussion of suburban social systems and the second was a comparison of postures toward mass society in three different suburbs viewed against the background of a small town. Theorizing at a highly abstract level usually evokes a lingering sense of guilt over the differentiating features that must be temporarily ignored. And, vice versa, discussions of the distinctive features are haunted by the impulse toward underlying generalities. There is no formula for avoiding the twin hazards of excessive generality and excessive specificity other than cultivating responsiveness to the conscience pangs aroused by excess in either direction.

Rather than concluding with any final formulations about American community life, it seems appropriate to end by again mentioning two perspectives. The first is constituted by the series of community studies summarized and interpreted in this volume. It serves to identify the historical processes shaping American community life and points to an outcome summarized in the conception of mass society. Paradoxically, this conception is most useful when it is pushed downward to lower levels of abstraction—mass society in its suburban embodiment or mass society in its small-town embodiment. The second perspective attempts to further "humanize" the first. No matter how tremendous the pressure to treat oneself and others as objects in mass society, vital human dramas are still enacted. Anthropology and psychoanalysis help to illuminate these dramas, as do the traditional "humanistic" disciplines. It is the special responsibility of the community study to keep these dramas continually in focus. No matter how far from the center of the stage, they do provide a meeting ground for people within a community which the depersonalizing forces of mass society can diminish but never destroy. There is even a possibility that sociologists will contribute to the enlargement of these meeting grounds, but this can happen only if we keep reminding ourselves that we are studying *human* communities and if we mold our theories and methods accordingly.

EPILOGUE

I. Park, the Lynds, and Varieties of Imaginative
Community Sociology

This entire book has established a perspective for treating community studies as case studies showing the workings of fundamental social processes in specific American contexts. It has demonstrated that sociological reports on communities contain vital information about broad American social changes and established a framework for initiating further studies of this kind. It is clear, however, that there are different styles of community sociology. At first glance, Park and the Lynds would seem to be poles apart in their intellectual orientations and their research styles. They actually are in many respects. But it must also be noted that they share a concern with historically significant processes and problems which entitles both of them to consideration as exemplars of what C. Wright Mills has aptly called "the sociological imagination."[1]

Community sociologists have a fine model of imaginative sociology always at their fingertips: the example provided by Robert Park. Park's collected essays demonstrate the way in which he was able to transform passing observations of the urban scene into important social insights. His insatiable curiosity was supported by dogged persistence at trying to grasp the hidden meanings of life in alien urban worlds. He rarely succumbed to the temptation to translate these alien meanings into familiar terms before he had thoroughly understood them as they were understood by the persons originally

[1] I want gratefully to acknowledge the assistance provided by the following recent work in helping to shape my orientation toward sociological theory and research: C. Wright Mills' book *The Sociological Imagination*, Barrington Moore's chapter on the strategy of social science in his *Political Power and Social Theory*, David Riesman's observations on social science research in *Individualism Reconsidered*, and a forthcoming essay titled "Social Theory and Social Research" by Joseph Bensman and Arthur Vidich. I would also like to thank David Rogers, Jay Schulman, and Mason Griff for discussing these topics with me.

holding them. William James's classic essay "On a Certain Blindness in Human Beings"[2] served as a benchmark for Park. The main point of this essay, and perhaps the main point of James's approach to psychology, was the injunction that the tendency to categorize experience prematurely, and particularly, to categorize prematurely the meaning of other people's experience, is the greatest threat to full human existence. On the other hand, openness to fresh experience and empathy with the alien meanings of other people's experience offer the most promising basis for creative life and thought both in social science and in the everyday world. This was the central insight that Park learned from William James and advocated throughout his life.

There are many other sources for this insight. A psychoanalytic version of it presented by Ernest Schachtel was discussed in Chapter 11. It seems to be the common property of contemporary phenomenologists, existentialists, some psychoanalysts, and several creative artists. Georg Simmel, who also influenced Park, had a profound intuitive grasp of cultural meanings which rendered his formal sociological essays far more interesting than parallel efforts by less sensitive formalists. Max Weber's emphasis on "Verstehen" as the basis for sociological interpretation points in the same direction. The problem of alien meanings was taken up directly in Chapters 10 and 11 with reference to selected psychoanalytic and anthropological theorists, but it actually appeared in every chapter in this book since the task of the field worker in each instance was penetrating and portraying these hidden meanings. This conception has many implications for social science, some of which will be taken up in this epilogue. *However it is important to underscore at the outset the fact that it is a terribly easy conception to propound but an incredibly difficult*

[2] This essay appears in James's *Talks on Psychology and Life's Ideals*, New York, Henry Holt, 1901.

[305]

conception to appropriate and integrate into one's working equipment.

In any event, the concept helped Park to remain interesting throughout his lengthy career. Park's strategy involved other elements besides persistent curiosity. As a scientist he recognized the need to classify and interpret the ever-changing urban world. He was continually evolving schemes for doing so but, unlike some contemporary theorists, he managed to remain sufficiently detached from his schemes to retain fruitful contact with changing realities. One might well inquire how he succeeded in holding the elements in the scientific enterprise in such suspension. The best answer is to look very closely at his essays. There is no rhetoric about the need for politically significant research, or calls for linking research to theory, or any of the other slogan-crutches that most sociologists now depend upon. He did not need the crutches because he had a deep sense of how to go about accomplishing the movements that we find so difficult. Our crutches all too often only prolong rather than mitigate our disorder.

Some more practical directives can be gleaned from Park's example. He seems to have consistently refused to settle down to a single level of sociological analysis. Looking over his collected papers, one finds him ranging from essays rooted in studies of the city of Chicago with its slums and ghettos to broad essays on the city and civilization. His interests in race move from the etiquette of race relations in the South through the career of the Africans in Brazil to general discussions of the relation between cultural conflict, personality, and marginality. The foregoing are selected from chapter titles in *Human Communities* and *Race and Culture*. The volume called *Society* includes everything from human nature, collective behavior, and social planning to strikes, revolution, war and politics, the natural history of the newspaper, industrial fatigue, and finally a synthetic essay on sociology and the social sciences.

What can we learn from this display of Park's free-ranging mentality and his romantic temper? His vision of the sociological enterprise was highly journalistic in the best sense. He turned to the burning issues of his time and wrote passionately about them. But he also wrote scientifically. The substantive essays are very frequently introductions to empirical studies undertaken by his students in which they reported on slums, ghettos, and strikes which they had actually observed as participants or through documentary sources, to the best of their abilities. The essays demonstrate Park's capacity to merge their perceptions with his own in order to highlight significant generalities amid the welter of detail.

But there is more unity and even system in Park's work than focusing on the enormous range of his essays might suggest. From his early work onward, he was always struggling with the need to cast his observations on city life into systematic form. His sense of reality saved him from freezing any of the many levels of systematization in his work into final form. The first chapter in this book tries to reopen one of his systematic approaches—his theory of natural areas. He himself seems to have had greater confidence in the potentialities of systematic ecology, but he never allowed this confidence to blind him to points at which the requirements of ecological systematization were rendering distinctively sociological problems and processes more rather than less obscure. Park's essay "The Urban Community as a Spatial Pattern and Moral Order"[3] shows his almost unique capacity to hold sociology and ecology separate, while at the same time extracting the maximum significance from their legitimate interrelationships. His several essays on human ecology all display this quality.

Park's distinctive accomplishment could be defined as keeping his systematic theorizing attuned to his exploration of emerging social patterns and problems, without allowing the

[3] Park's essay is reprinted as Chapter 14 in his *Human Communities*.

messages from one source to drown out the other. He was constantly revising his systematic categories to improve their interpretive powers and he was always willing to explore a problem area that seemed to defy these categories. Problems and interpretations always took precedence over concepts and systems. To the end of his life he remained interested in the world about him, and partly because of that he remains interesting to us.

There is another way in which community sociologists present us with useful observations. This is the path followed by the Lynds, and by most authors of studies of specific communities. It differs considerably from Park's broad theoretical perspective and depends upon rather different kinds of sociological skills. With all of his gifts for stimulating students to get out into the city to do research, Park himself never did very much sustained work of this kind. He provided the overview and linked detailed studies undertaken by his students to broader interpretations of urban processes. He must have been a fine "casual" observer, but not the kind of person who moves into a community, lives there patiently for several years, and then departs to write a book about it.

The Lynds did just that. They and their research team moved to Muncie, lived there for a year, and departed to write their objective report. One aspect of the popularity of *Middletown* at the time of its publication and for several years thereafter was its privileged status as one of the first books in which social scientists report on an *expedition* to an American town. Treating an American town as one would a primitive community had an odd appeal since it promised to combine inside and outside perspectives. Readers were prepared for such a combination by the novels of Sinclair Lewis. But this inside-dopester appeal went even further. People in the cities were able to assume, not only that they knew what went on in this Midwest town, but that they somehow knew it better than the people who lived there. We still share some of that feeling today. Not only do we know more about Muncie in the twenties and thirties than did the participants in community affairs during those years, but we know more about Crestwood Heights than the suburbanites swept along by its social system, and we certainly know a lot more about Springdale than any of its frightened citizens are able to allow themselves to find out.

There is an irresistible tendency for the reader of a community study to assume an all-knowing attitude toward the community he is reading about. The Lynds succeed in creating this attitude to an unusual extent. One literary technique they use for accomplishing this is a kind of dead-pan irony through which they report what the people of Muncie are experiencing, while simultaneously critically commenting on that experience. Compared with Sinclair Lewis's cutting use of this same technique in his novels or with Thorstein Veblen's savage ironic manner, the Lynds are distinguished by their restraint, but their technique is related to these other forms of irony.

Irony is usually employed by the Lynds when aspects of conspicuous consumption are described and it serves to highlight the obsessiveness of this quest for commodities, as well as the human sacrifices that it occasions. It is used to underscore deftly the confusion about the meaning of their experiences that these early "status seekers" display in their everyday lives. The continual use of contrasting images is one of the Lynds' standard devices. The authors describe a pattern from Middletown of 1890, emphasizing its relationships to a coherent system of community arrangements. They then juxtapose a pattern from Middletown of 1924, showing the way in which intrinsic meanings have been sacrificed to the pursuit of an ever-rising standard of living. The total effect is probably more devastating than the direct ironies of Sinclair Lewis or Thorstein Veblen because the stylistic mechanisms remain unnoticed. Concealed within a disguise of scientific objectivity is a lasting object lesson in the futilities of conspicuous consumption.

Naturally, the Lynds also offer a great deal more. But this one element is worth highlighting because it casts light on a dimension of the community study which is usually neglected in sociological discussions of research reports or methodology: the vital way in which the style of the report affects the scientific quality of the research. Short of reading the field notes, no one can know what the primary observations may have been, and so the published study becomes the sole avenue that the general public or other scientific readers have to the community in question. It is surprising that so little attention has been paid to the crucial role played by reportorial style, since it is at least as important as the more frequently noticed style for conducting the field work which does occasionally find its way into methodological appendices.

Surveying many community studies in an effort at finding general patterns forced attention to the ineffaceable stylistic elements in the field work and in the research report. The management of both these elements helps to distinguish the good study from the inferior one, but very few clues as to how they are to be managed appear in discussions of research methodology. There are two aspects of the situation. In his field work, the community sociologist is limited to gathering those meanings which participants are willing to divulge to the kind of person he presents himself as being, and which fall within his empathic range. Secondly, regardless of his research artistry, the scientific and general public can get access only to that segment of his findings which he can crystallize through the written word.

The first aspect is, of course, most obvious. Thus, the fact that the research team of *Deep South* included a Negro family and a white family meant that kinds of data were communicated to the field workers that would have been extremely difficult for a single fieldworker of either race or sex to obtain.

Similarly, John Dollard's psychoanalytic training and orientation led him to observe and report signs of unconscious processes such as dreams, slips, and free associations that a differently trained observer would never have noticed. Equally
important, it allowed him to interpret Southern patterns that
someone without psychoanalytic training could never have
dealt with. These orientations are an integral element in the
structure of the study and must be taken into account if the
findings are to be fully understood.

This aspect of the situation—the effects of observer styles
on the observational process and products—is all too easily
reduced to a topic like "interviewer effects," with the assumption that it can be solved by sending out different kinds of in-
terviewers and noting where they influence the data in order to
control the effects of discrepancies. This may be possible with
some survey research projects, but it does not help the community sociologist who is dependent upon lasting structured
relations with his respondents to get access to private meanings
of which they themselves may be only momentarily aware. A
closer approximation to the kind of methodology needed is
contained in the literature on participant-observation, but
anyone who has ever tried this difficult technique knows that
the skills of participant-observation resist codification.

This is not meant to discourage efforts at codifying participant-observation; it rather aims at reorienting them within a
different context. Important steps in this new direction have
already been taken. Two articles in the *American Journal of
Sociology* for January 1955, "Problems in Participant Observation," by Morris S. Schwartz and Charlotte Green Schwartz,
and "Participant Observation and the Collection and Interpretation of Data," by Arthur Vidich, deserve special mention.
The Schwartzes analyze the crucial role played by the way in
which the observer manages his own anxiety in determining the kind of data he collects. Vidich interprets the complex
interplay between the observer's changing involvement as a

study progresses, the images that respondents come to hold of him, and his capacity to look objectively at these images and at the community under observation. These kinds of codification are exceptionally useful because they aim at elucidating the role played by specific subjective variables to bring them under better control. Being forewarned is to be forearmed, and fieldworkers can help each other immensely by sharing experiences in this fashion. Needless to say, creative responses to the anxieties aroused by playing the role of participant-observer of the sort that achieve genuine self-insight and original perspectives on the groups studied can be encouraged, but never guaranteed, by methodological forewarnings.

The roles played by personal styles in conducting research and reporting findings deserve far more attention than they ordinarily receive. Looking at *Middletown* with this in mind, we can see that the Lynds employed somewhat different research styles in getting information from members of the two classes. This research artistry enabled them to penetrate the distinctive worlds in which each class lived without losing their grasp of the objective structure of the community. Their reportorial skill allowed them to portray both working class and business class images of Muncie in rich detail, retaining the full flavor of genuine inside perspectives. Dramatic irony helped them to do this by providing a framework in which the integral perspectives of the two classes could be set against a third objective perspective in a balanced fashion. The Lynds were judging Muncie while describing it, but their style for doing so entailed a complex appreciation of the dramatic ironies of human life. They never substituted their interpretation of local affairs for the interpretations held by the two classes, but they also never allowed their own interpretation to slip completely out of sight. Instead, they deployed it as a rational backdrop to highlight irrational overtones in the perspectives of both classes. They did this so tactfully that people deeply committed to the values of industrial society were able to enjoy the book despite its ironic commentary on these values.

Later studies have only occasionally been able to attain the same blend of critical detachment and discerning sympathy that the Lynds achieved. Indeed, the Lynds themselves were driven to adopt a much sharper tone in *Middletown in Transition*. Apparently their exasperation with the community's failure to face up to what the authors felt were the realities of the Great Depression left them incapable of treating Muncie with the same gentle irony displayed earlier. The Lynds left a permanent mark on the sociological imagination. They showed that it was possible for sophisticated intellectuals to root themselves in a community that was alien to their present lives, if not to their past experiences, sufficiently to produce a study of that community which conveyed the inner meanings of its round of life vividly and truthfully. They evolved a descriptive format for commenting with mild irony on the hitherto unnoticed dramas of everyday life so that these achieved significance. They wrote the miniature epic of this Midwest town and encouraged other sociological storytellers to try their hands at doing the same for other American communities.

II. Dramatic Irony, Irrationality, and the Role of the Observer

A unified conception of community sociology must draw upon the example of the explicit theorist set by Park and upon the example of the perceptive observer set by the Lynds. Further development of the field rests upon syntheses of these two styles. Certain peculiar difficulties arise when studying contemporary American communities, especially in connection with the need for controlling field observations and theorizing to be certain that they continue to focus on significant problems. These difficulties sometimes stem from the research orientations of the field worker; there is a tendency to find only what the techniques that the sociologist has been trained to use permit him to find. Many of the most popular techniques drawn from contemporary survey research are not too helpful in community studies. Theoretical orientations can also become sources of difficulty. The prevalent emphasis on systematic theories is poorly suited to guide studies concerned with settings displaying the broad range of variation found in American communities. The danger here is that the field worker will tailor his findings to fit his conceptual categories. These hazards are present in any scientific endeavor, but they are multiplied when the observer is prone to become personally involved with the problems of the group he is studying. Such involvement can easily become threatening, and standardized research techniques, as well as elaborate logical theories, can become defenses against recognizing the real inconsistencies and real problems of the subjects. This is most likely to happen when there are close resemblances between the life problems of the observer and the social problems of the community that is being studied.

Most social scientists tend to have high regard for objective insight and, correspondingly, tend to feel little sympathy with people who interpret their own behavior in ideologically confused terms. Thus the Lynds lost a good deal of their sympathy for the people of Muncie when they found them clinging to their old symbols despite changing realities wrought by the depression. This emphasis on rational understanding of social processes is an important factor motivating people to study sociology.

A sociologist committed to rationality, who is confronted by behavior on the part of another person that he can show to be objectively different from and contradictory to what the person himself thinks he is doing, is likely to become emotionally upset. There are many ways in which this emotional response can be handled. Marxists and some students of Karl Mannheim would define the person's incorrect explanation for his behavior as an "ideology," and concern themselves with ways of bringing this misguided person around to "true consciousness" by making him aware of the "real" state of affairs. Freudians locate the source of the unreasonable explanation in the person's (usually reclassified as a patient) deeper psychic needs, at least some of which are "unconscious" and which cannot be brought to consciousness without extensive psychotherapy. Both Marxists and Freudians respond to demonstrable "irrationalities" by explaining them in terms of forces beyond the threshold of the patient's awareness and then insist upon the importance of overcoming "resistances" so that the person (patient, worker, etc.) can be made fully aware of his situation.

Community sociologists confront this crisis on a somewhat less abstract level than most social theorists. Each generation meets it in a different form, depending upon the kind of commitment to theories of objective structures that it holds and the extent to which these theories diverge from common-sense interpretations of the social worlds studied. Thus, the Lynds, perhaps somewhat under the influence of Marxism and prodded by the depression in the thirties, found themselves growing more critical of local irrationalities than they had been in the twenties.

Southern communities would disturb most Northern sociologists. From the Northern liberal standpoint, which is likely to coincide with the definition of objective reality that the sociologist adopts, the Southern caste system is "crazy" and the justifications for it are little more than "insane" rationalizations. John Dollard's *Caste and Class in a Southern Town* skillfully employs both Marxist and Freudian concepts to unmask the effects of these irrationalities at many levels. Allison Davis and the Gardners show more restraint in *Deep South*, although their indignation breaks through at several points. It is one thing to theorize about social and psychic irrationalities and another thing to watch them destroy human lives before your eyes.

The entire world was shocked by an outbreak of irrationality in the thirties that nearly defied comprehension in objective terms. That was, of course, the unbelievable depredations committed by ordinary Germans when they became Nazis. It is an enduring testimonial to the life-saving power of rational thought that the best interpretations of this phenomenon were produced by German refugee scholars such as Franz Neumann, whose *Behemoth* first appeared in 1942, and Erich Fromm, whose *Escape from Freedom* was published in 1941. The other great disillusioning experience of the thirties and forties, apart from the depression mentioned earlier, was of course the exposure of totalitarian Stalinism in the Soviet Union. These three crises—the loss of faith in capitalism during the depression, the loss of faith in Marxism and Stalinism, and the shock of Nazism—form the background for the present situation of social theorizing and social research.

A book much larger than this one would be needed to trace the effects of these upheavals on American sociology. Yet they must at least be mentioned if the present situation is to be understood. Part II of this book traces some of the effects of the depression and World War II on the communities studied. Attention is called to their effects on the sociologists doing the studies. Thus, William Foote Whyte's very sympathetic attitude toward Italian working-class values was encouraged by the climate of the period as much as by his personal proclivities. It is doubtful that any sociologist in the fifties would be drawn to working-class or ethnic communities in the same fashion.

It is not surprising that community sociologists in the fifties should turn to suburbia. There are usually varieties of the species located conveniently near the academic base of operations, if indeed, they are not the home community of the investigator. Middle-class sociologists find it easy to establish rapport with middle-class suburbanites. The insatiable curiosity about alien sub-cultures and the willingness to devote long years to comprehending them, which Park's students displayed, seems to have shifted over to anthropology if indeed it still exists at all. But the focus on suburbia has its own hazards.

As our discussions of Park Forest, Crestwood Heights, and Exurbia suggest, the irrationalities of suburbia are by no means insignificant. More important, they verge very close to the commonplace values and problems of the middle-class sociologist. After all, the Organization Men of Park Forest include Organization Scientists with their sub-type, the Organization Social Scientist. The desperate struggles of the would-be creative artist in suburbia to sustain his secret dream have their parallel in the struggles of the dedicated teacher who finds himself doing trivial research in order to achieve a satisfactory publication record and comply with the demands of the academic marketplace. Crestwood Heights, with its patent contradiction between "success" values and "maturity" values, its women caught between homemaking and careers, its schools squeezed between "progressive" ideologies and "competitive" realities, and its crew of psychiatric repairmen is itself part of the characteristic experience of many intellectuals. Those who have not yet arrived at this point are probably struggling to do so.

Journalist-sociologists like Whyte and Spectorsky probably had less difficulty establishing proper distance from the suburbs they studied than did the clinically oriented sociologists who studied Crestwood Heights. The fact that Seeley and his associates were involved both as therapist-experts and as sociological investigators probably created difficulties for them. The reportorial style of *Crestwood Heights* reveals considerable ambivalence. The authors tend to write long complicated sentences, one-half of which will reflect the community's favorable self-interpretation, while the other half conveys the investigator's critical feeling. Spectorsky maintained a consistently cynical attitude toward Exurbia, while still remaining fairly accurate. William H. Whyte created a style which combined a crusading journalist's critical zeal with the sympathetic understanding of a good social scientist.

It is the intellectual's conviction that life holds more important purposes than the pursuit of economic success that disposes him critically towards the life goals of suburbanites. Research only aggravates this disposition if it reveals that other professed suburban goals really conceal acquisitive purposes or serve to ease some of the strains that the success drive generates. Very soon, the whole life routine of the suburb begins to look like a massive charade in which everyone plays at appearing to be rounded human beings while dedicating their energies to pursuing status symbols and career opportunities. Intellectuals have a hard time appreciating the extent to which the quest for success can become the sole meaning of life, simply because we badly want to avoid recognizing the extent to which it has displaced other meanings in our own lives. We are disturbed by the suburban showcase home because our own homes can so easily become showcases differing only in the kind of objects that we put on display.

During the twenties the Lynds were able to achieve greater distance from the spread of acquisitive motives and values because they felt that the dominance of these values might be temporary and hoped that the publication of their book would hasten public enlightenment. Even during the thirties, it was still possible to hope that the "false consciousness" displayed by Muncie workers was confined to smaller towns, while the workers in cities were becoming aware of the contradictions of capitalism and the need for socialism. Referring back to the historical context mentioned earlier, we no longer have general theories to justify any hope that this situation, in which acquisition and display have been transformed from means into ends, will be modified by historical forces. In the light of the apparent alternatives, we are not even certain that we would want it to be.

But there is still an underground resistance to the acceptance

[318]

of affluence as the highest goal of human life. One expression of this resistance is fascination with the human distortions and internal contradictions appearing in suburbia. If the dramas of suburban life can be held at a distance, then their limitations become apparent. Yet the sociologist who studies this condition exposes himself to greater hazards than his predecessors. He can no longer fall back upon the march of history, and he must accept the struggles of his subjects as part of his own experience.

Each community sociologist is left to work out his own relationship to the community that he studies and the best clues to his solution are contained in the style of the final report. The omnipotent feeling on the part of the reader of a community study can only be a pale reflection of the omnipotence felt by the author as he literally creates the picture of the community while he writes his book. Naturally, this creative process involves exploring the irrational self-images and community images held by his subjects, along with the objective structures that his scientific framework and observational stance enable him to discern. Finally, what he finds and what he reports is determined as much by his sympathetic and experiential limits as by anything else. *The quality of the study hinges largely upon his capacity to broaden these limits so as to comprehend human behavior which expresses meanings that he ordinarily would not entertain in his personal world. In doing and synthesizing the study, he dissolves the boundaries of his old self and recreates new boundaries simultaneously with his creation of a new and more accurate image of the community.* This is the reason why the sociologist has to choose his community very carefully. It is likely to shape him just as much as he shapes it; either way, the two are inseparably wedded for the duration of the research and probably for a long time thereafter.

Acknowledging the subjective elements in social science theorizing and research should not be construed as a defense of irrationality. These subjective influences are operative whether we recognize them or not. The only way in which their influence can be controlled, and reasonable standards of objectivity maintained, is through the development of methods for making subjective responses accessible to objective scrutiny during the several phases of the scientific process. This problem can be approached from many angles, only a few of which can be taken up here. Sociologists have usually been willing to admit the involvement of values in the selection of problems to be studied. They have been far less willing to admit that the choice of concepts, research techniques, and reportorial styles are all equally value-laden choices. In some circles, the mere suggestion that emotional influences color the scientific process from beginning to end is sufficient to provoke charges of heresy.

Several recent contributions in different fields raise many of these questions in a helpful fashion and they provide necessary background for the treatment of community sociology advanced in this book. H. Stuart Hughes's *Consciousness and Society* is a fundamental source for exploring the different ways in which several European thinkers writing between 1890 and 1920 simultaneously discovered the importance of the subjective responses and commitments of the observer in shaping the character of the observations that he was able to make. He places this discovery in its context as a reaction to positivism and to Marxism. The thinkers cited include philosophers, historians, sociologists, psychologists, and creative artists. They share recognition of "perspectivism" in one of its many versions. Hughes's contribution lies in showing relationships between such diverse figures as Max Weber, Sigmund Freud, George Sorel, Benedetto Croce, Thomas Mann, and Luigi Pirandello among others. He never clearly distinguishes between the various facets of this discovery, but identifying the generation and interpreting its significant contributions was itself a huge task. Further work in this area will have to start with Hughes's book.

Karl Mannheim's many essays on the sociology of knowledge contain several specific leads for applying "perspectivism" or "historicism" to the scientific process. His chapters on American sociology and on German sociology in *Essays on Sociology and Social Psychology* demonstrate his skill at using the sociology of knowledge to illuminate the intellectual commitments of working social scientists in a fashion that reveals hidden strengths and weaknesses. He shows how these commitments are shaped by the social climate and experiences of intellectuals in each country. The use of the sociology of knowledge as an instrument for exploring and improving the work styles of sociologists has not yet become an integral part of the working equipment of American practitioners. The most interesting recent effort along this line is found in C. Wright Mills's *The Sociological Imagination*.

However, if Hughes and Mannheim are correct, general interpretations of the situation of the intellectual or even specific interpretations of the pressures within a single discipline are hardly sufficient. We need much more specific directives capable of being deployed in field situations to correct biases and redirect attention while social observation is in progress. The only social science to develop a full-scale apparatus for doing this is psychoanalysis. And that apparatus detects and channels emotional biases without bringing intellectual commitments or the effects of social position to awareness.

There is only one work on methodology that tries to tackle this problem in specific terms. That is *The Study of Human Nature*, by David Lindsey Watson. This book actually analyzes the ways in which emotional and intellectual attachments affect the work of the social scientist. It has received almost no attention within the social sciences, a neglect which probably stems in part from the author's refusal to pull any punches

[321]

or to hide behind a neutral vocabulary. A recent book, *Personal Knowledge*, by Michael Polanyi, explores the general philosophical and psychological underpinnings of this problem to show how the scientist's commitments give form to his data and to his theorizing. Still another book, this time from a philosopher of the "Wittgensteinian" persuasion, *The Idea of a Social Science*, by Peter Winch, restates the case for meaningful interpretations in compelling terms. Perhaps these latter books will provide impetus for reexamining the heresies propounded by Watson. Philosophers of history and working historians have long been concerned with the role played by the intellectual climate and capacities of the observer in shaping his endeavors. Sociologists can still learn a good deal from the writings of R. G. Collingwood, Benedetto Croce, and Ortega y Gasset on this subject.[4]

But neither the philosophy of history nor the sociology of knowledge has yet succeeded in matching psychoanalysis with respect to the development of working instruments for exploring subjective influences. Analytic concepts of "free floating attention," or, as Reik puts it, "listening with the third ear," describe an attitude toward observation which synthesizes self-awareness and objective scrutiny. The conception of "counter-transference," around which so much of analytic training revolves, is the key to systematic control over observer influences. In this respect, psychoanalysis is far in advance of other schools of psychology and, indeed, of the other social sciences, in achieving true objectivity.[5] Sociologists have much

---

[4] Books of particular importance on his theme include R. G. Collingwood's *The Idea of History*, Benedetto Croce's *History as the Story of Liberty*, and Ortega y Gasset's *Toward a Philosophy of History*. See also the excellent collections, *Varieties of History*, by Fritz Stern, and *The Philosophy of History in Our Time*, by Hans Meyerhoff.

[5] Theodor Reik's contributions are scattered throughout his numerous books, but his collection *The Search Within* contains material especially relevant to this discussion. Psychoanalytic conceptions of transference and counter-transference are explored from a Sullivanian standpoint in Benjamin Wolstein's useful book, *Transference*. In addition to the theorists mentioned in Chapter 11, the writings of Lawrence Kubie and

to learn from psychoanalysis with respect to developing techniques for controlling the influence of emotional variables on the research process. The new orientation to participant-observation by the Schwartzes and Vidich mentioned earlier has begun to take these lessons into account.

The full implications for the whole field of sociology that this rediscovery of the subjective element holds will take a long time to unfold. We have barely begun to digest the work of the theorists discussed by H. Stuart Hughes and there are new forms of positivism which threaten to distort the work of this earlier generation beyond recognition while claiming to be its legitimate offspring. Community sociologists are fortunate in having the examples of Park and the Lynds to demonstrate the value of quite different but equally effective personal styles as agencies for organizing and interpreting community processes. This book takes a step in the direction of explicating the elements that made these personal styles so effective, and therefore rendering them more susceptible to objective scrutiny.

There is a frame of reference emerging within sociology that promises to combine many of the insights of "perspectivism" with more practical research orientations. This frame of reference is named "dramaturgic," "dramatistic," or simply dramatic by its originators.[6] We have employed it throughout

---

Paul Goodman helped to shape the psychoanalytic conceptions adopted in this book. Kubie's *Neurotic Distortion of the Creative Process* and Goodman's *Gestalt Therapy* (co-authored with Frederick Perls and Ralph Hefferline) offer many original perspectives on Freudian theory and therapy.

[6] A number of critics of literature, drama, and the arts have contributed importantly to the development of the dramatic frame of reference. Kenneth Burke is especially significant and all of his books, ranging from the early *Permanence and Change* to his recent *A Rhetoric of Motives*, deal with aspects of the problem. Other books that I have found useful include Francis Fergusson's *Idea of a Theatre*, William Empson's *Some Versions of Pastoral*, Robert Blackmur's *Language as Gesture*, Robert Langbaum's *The Poetry of Experience*, and Harold Rosenberg's *Tradition of the New*. No sociological effort at formulating a dramatic theory can ignore these vital contributions.

[323]

this book and it is finally time to acknowledge our debt to Everett Hughes, Erving Goffman, and Anselm Strauss, who are among the leading contributors to the development of a dramatic frame of reference in American sociology. As with "perspectivism," the intellectual antecedents of the movement can be only summarily mentioned. Unfortunately, we have no study comparable to H. Stuart Hughes's book to summarize this tradition, but it is a tradition that grew out of the Chicago school of urban sociologists so its character will be easily grasped by anyone who has followed the argument of this book thus far.

One seminal theorist in this tradition was George Herbert Mead, whose *Mind, Self and Society* nourished an entire generation of social psychologists. Mead was a colleague of Parks and clearly influenced Park's conception of social psychological processes at many points. Everett Hughes has come closest to achieving a synthesis of Mead's study of symbolic processes with Park's interest in urban patterns. The recent publication of Everett Hughes's essays, *Men and Their Work*, added to his earlier collection on race, *Where Peoples Meet*, permit someone not trained by him at Chicago to appreciate the scope of his contribution. Adopting the same essay form employed so skillfully by Park, Everett Hughes offers a panoramic perspective on many important aspects of modern social life. He turned from cities to occupations as his main subject of inquiry but he carried over Park's capacity to appreciate alien meanings and accord them the dignity they deserve.

Like Park, Everett Hughes seems to avoid developing comprehensive systems and treats his concepts as tools for inquiry without being too concerned about logical interconnections or levels of abstraction. Again like Park, his students have begun to formulate their teacher's sociological style more systematically than he seems to have cared to. So, we find Erving Goffman presenting a close-knit dramaturgic framework in *The Presentation of Self in Everyday Life* which permits the exploration of social settings as back and front regions in which performers control the impressions they make on audiences. Goffman's concepts point at hidden dramatic structures and processes with enviable precision, while Anselm Strauss in *Mirrors and Masks* shows how identity transformations are accomplished in diverse institutional and cultural settings.

The frame of reference being developed by Everett Hughes, Strauss, and Goffman is obviously applicable to the field of community studies. It provides a rationale for the employment of dramatic irony which appeared as a central element in the Lynds' style. From a dramatic standpoint, the central problem of the community sociologist is to achieve an objective perspective that encompasses the partial perspectives held by various groups in the community in such a fashion as to call attention to hidden processes without losing sight of the meanings of the various partial perspectives. The playwright seeks to present his characters sympathetically without going so far as to allow the sympathy they evoke to swallow the larger meanings that emerge when they are viewed within the context of the entire plot and action of the play. The play suffers as much when sympathy with a single character obscures the context as when the context is allowed to override full presentation of diverse characters. The playwright seeks a profound balance and it is similar to the balance sought by the community sociologist.

Dramatic sensibility then consists of the capacity to encompass multiple interpretations of a social world within a larger context which distinguishes objective structures without obliterating subjective meanings. This is essentially the problem of the participant-observer discussed earlier. He must be capable of moving close to and away from his informants so that he grasps their point of view while, at the same time, exploring its relationship to his *evolving composite* picture of the community. As Vidich notes in his essay on participant-observation, maintaining the proper balance between

[325]

detachment and involvement confronts the sociologist with different problems at early, intermediate, and final stages of a research project. *Small Town in Mass Society* demonstrates that Vidich and Bensman solved these problems beautifully despite the special difficulties that confronted them. Springdale could hardly have been studied without dramatic distinctions and sensibilities since the whole life of the town revolved around public performances which contradicted objective realities. One can only admire the authors' capacity to describe the details of the public ceremonies, conveying so much of their inner meaning despite their constant awareness that the ceremonies falsified the main objective political and economic realities of small-town life. In Springdale, the contradictions were so sharp that Vidich and Bensman did not need ironic tones to convey their findings. Instead, they adopted a clinical style which somehow still captured the inner overtones of local life. These overtones consisted of endless clichés for the most part, but if this is what the field workers find, then it is what they are obliged to report.

There is a strong possibility that team research is especially effective in community studies because it broadens the dramatic range of the members. It is worth noting that teamwork appeared in community studies long before it became fashionable in social science as a whole. Having more than one person doing the study means that different avenues into the community can be established and various perspectives brought into focus that might escape a single investigator. It creates a situation in which team members can control each others' tendencies toward overinvolvement or overdetachment while the research is in progress and while the report is being written. William Foote Whyte, in his appendix to the second edition of *Street Corner Society*, shows us how a single investigator works through these problems. It would be extremely helpful if similar information could be obtained about the research process and the circumstances under which the final report was composed from such teams as the Lynds,

Warner and his associates, and Seeley's group, among others, before this vital information is permanently lost.

The dramatic perspective of the contemporary sociologist is not sustained by any definitive theories about the direction of historical processes, and it is bounded by terrible awareness of the destructive potentialities of irrational forces. We have learned a great deal about the ways in which human destinies are shaped by social and psychic forces beyond the conscious control of any single individuals. No student of human behavior today can afford to ignore the existence of many levels of consciousness in himself and his subjects, nor can he ignore the situational determinants of responses, including his own. The prospects for developing a fruitful dramatic theory in social science would seem to be highly promising, but the terrors of the irrational to which everyone has been exposed often tend to make a dehumanized behavioristic theory appear more attractive. The resonance of dubious images of physical science among social scientists has deeper roots than simply the desire to emulate a more popular group. When the social scientist concedes the decisive role of subjective influences in his work style and commits himself to some kind of dramaturgic perspective on the investigative process, he opens up a great many more problems than he solves. Sometimes this cannot be avoided and if the discoveries of the generation studied by H. Stuart Hughes have any validity the long way around may eventually prove more fruitful than the positivist short-cuts.

A full dramaturgic perspective on communities will have to include awareness of human dramas that receive comparatively little ceremonial celebration in contemporary American life. These have already been described in Chapters 10 and 11. They are, of course, the dramas connected with major phases of the life cycle. Viewed dispassionately, the status preoccupations and protective fictions found in great abundance in American communities hardly seem comparable to fullfledged irrationalities of modern totalitarian societies. The plight of suburbanites and small towners caught in self-defeating webs and rendered incapable of separating illusions from realities is still hardly comparable to that of either the victims or the perpetrators of totalitarian terror.

There is a sense, however, in which these minor irrationalities of an affluent society endanger human dignity at vital points. By weakening the capacity for rational social thought, they help to prepare the way for more serious irrational outbreaks if circumstances should change. Men who cannot separate fact from fiction in the local political struggles that go on around them will hardly be able to make effective decisions during a real national crisis. The very fabric of democratic society is weakened when men can no longer understand their everyday worlds. *The implications of these minor irrationalities loom even larger when they are measured against Enlightenment faith in human rationality or against Renaissance images of human possibilities.* These are the higher dramatic sources, the full measures of tragedy and comedy, against which the community sociologist must assess the quality of the everyday lives of the people he studies. The highest standards of human achievement must be invoked at crucial points to govern our dramatic perceptions. Anything less would mean cheating our subjects and ourselves by neglecting the great images of man in the Western tradition in favor of a stripped-down mechanical model. History may no longer be on our side, but if we can free ourselves from all coercions to focus exclusively on the present, perhaps we can yet use history to help us visualize future possibilities in the light cast by the peak achievements and perspectives of the past.[7]

Since this is almost the end of the book, it may prove helpful to make explicit some of the implications of the title.

[7] For further discussion of these uses of history, see the introduction, "Identity and History," and the essays in Part III, in *Identity and Anxiety*, edited by Stein, Vidich, and White. See also Herbert Muller's excellent study, *The Uses of the Past.*

Viewed through the studies taken up in Parts I and II, American communities can be seen continuing the vital processes uncovered in Muncie by the Lynds. Substantive values and traditional patterns are continually being discarded or elevated to fictional status whenever they threaten the pursuit of commodities or careers. Community ties become increasingly dispensable, finally extending even into the nuclear family, and we are forced to watch children dispensing with their parents at an ever earlier age in suburbia. The process seems to have two poles. On the one hand, individuals become increasingly dependent upon centralized authorities and agencies in all areas of life. On the other, personal loyalties decrease their range with the successive weakening of national ties, regional ties, community ties, neighborhood ties, family ties and finally, ties to a coherent image of one's self. These polar processes of heightened functional dependence and diminished loyalties appear in most sociological diagnoses of our time. However, we have only recently become aware of the full extent of human vulnerability and manipulability. We live in a period when the "existentialist" experience, the feeling of total "shipwreck," is no longer the exclusive prerogative of extraordinarily sensitive poets and philosophers. Instead, it has become the last shared experience, touching everyone in the whole society although only a few are able to express it effectively.

Suburbia is so fascinating just because it reveals the "eclipse" of community at one of its darkest moments while still hinting at the light that may follow. Here the endless disguises and contortions imposed by the career appear to form a giant web enveloping the trapped participants despite their continual movement from one side to the other. But here, too, the problem of identity finds its most lucid expression. The struggle for maturity in Crestwood Heights, for secure roots in Park Forest, and creativity in Exurbia, no matter how badly distorted, could become entering wedges for social change. These struggles reflect the desire for deeper human encounters and experiences

[329]

than those encouraged by the preoccupation with status, and they could become the occasion for identity transformations in which this preoccupation will assume a lesser place.

There is even an advantage in the extreme character of the modern plight. Having lost most traditional attachments, the individual is free to choose among a broader range of possibilities than ever before. The technological achievements of our society actually make alternatives more visible as well as more accessible than they are in most other societies. To take just a single instance, the profusion of attractively bound, inexpensive editions of literary, philosophic, and historical classics is both an opportunity and a hazard. The person trying to educate himself can easily be discouraged by the immense range of available materials but if he can overcome this initial discouragement he can own, study, and enjoy books which formerly would be found only in the largest libraries. Our very disillusionment with the present opens the way to sympathy with alien historical periods and cultures if the first reactions of apathy and confusion can only be overcome.

The community sociologist has to be constantly alert to efforts by the people he studies to rewrite their own scripts. Certainly, psychoanalysis represents one such effort. It aims at restoring openness and depth to one's memory schemata and experience schemata exactly as would primitive ritual dramas. It is a step toward the reappearance of genuine community, even if it be only a community-of-two constituted by the therapist and his patient. But the ceremony of psychotherapy, no matter how deep it may go, is still an initiation rite. The initiate must finally move into a larger communal framework to enjoy the dramatic possibilities that the ceremony reveals. Doing this at present presupposes sufficient imagination to write an original script, sufficient empathy to attract responsive fellow actors, and sufficient courage to refuse to play the many stereotyped roles that await everyone who walks onto the stages of modern life.

III. The Ideal Community Sociologist

Four components of the work of the community sociologist can be distinguished. Any specific practitioner is likely to concentrate on one or two, leaving the others relatively implicit in his work. It is possible that they require quite different kinds of talent and therefore the cultivation of some entails the sacrifice of others. For this reason they will be discussed here as if they constituted separate work styles, but always with the proviso that all four must enter into really useful theorizing and research. Community sociologists can emphasize ethnography, problem-solving, theorizing, or social philosophy in varying proportions, depending upon their training and talent.

The first style, the community ethnographer, is in low repute in American sociology today, though Park always had high regard for the urban ethnographer in his role as keeper of the historical record. This was especially plausible in the twenties because it appeared that many of the first-generation ethnic ghettos were doomed to extinction as their inhabitants left them to move to areas of second settlement where Americanized lifeways began to prevail. Getting a picture of life in these ghettos to capture the distinctive perspectives and experiences of transplanted marginal men seemed quite important. The incentive here was quite similar to the anthropologist's desire to record authentic primitive lifeways before these were modified by the influence of civilization. Any group's special experiences deserve a place in the record of human history; adding to this record is a worthy accomplishment.

Park's vision of the urban ethnographer went beyond this. He saw the sociologist serving an essential communicative function by furthering understanding between people who otherwise would remain alien to each other. In our huge impersonal cities, where most human encounters are mediated by superficial amenities, very few people are able to acquire familiarity with life in any sub-community besides the one in which they

[331]

happen to live. It could be argued that they do not necessarily have much objective knowledge about their own home territory since self-magnifying fictions are as prominent in urban neighborhoods and occupational groups as they are in small towns or suburbia.

It is easy to underestimate this aspect of the community study. Accurate and vivid description remains a vital element in any good study. No one who has read William Foote Whyte's *Street Corner Society* can continue to rest easy with the stereotyped image of the slum as an "impersonal jungle" nor can the reader of *Small Town in Mass Society* cling to nostalgic images of rural life such as might be inspired by Norman Rockwell's magazine covers. The author may well be able to describe a good many patterns that defy explanation at the time the study is written. By putting them on record, he makes it possible for later readers, familiar with the outcomes of the incipient processes that he studied, to develop more comprehensive interpretations than any observer on the scene at the time could have produced. Unlike much sociological writing, a community study always tells the reader substantive facts that he did not previously know, and for that reason it is invariably of some interest. Taken as a whole, the community studies of a particular epoch constitute a panorama revealing the varieties of ways of life that this period encouraged and allowed.

The image of the community sociologist as problem-solver also has to be explored on several levels. Certain kinds of community studies are undertaken at the behest of community leaders to gather information that these leaders consider essential for running local affairs. There is also a movement toward community "self-surveys" which presumably enable people to obtain a more objective picture of their own situations with an eye toward improving them. These ventures into "applied" sociology deserve encouragement, but their full potential will be realized only if lay participants and the participating sociologists are made aware of the resistances that real insight invariably must provoke, and tackle the difficult problem of devising techniques for overcoming these resistances. It is likely that some desired changes will prove beyond the control of local agencies while many others will challenge vested interests and cherished images to a point where little can be done to translate them into action.

Another style of problem-solving is linked to Park's conception of urban disorganization. Traditionally this has meant the study of slum conditions with an eye toward improving them. However, as suggested in Chapter 1, Park recognized deeper kinds of urban "troubles," many of them being direct concomitants of urban segregation and isolation. The anonymity of city life is always two-sided: on the one hand, it permits expression of individual talent; on the other, it leaves the individual vulnerable to complete isolation. Some urban sub-communities impose informal controls and obligatory standards of display which eliminate the possibility of creative marginality. We may have to plan for Bohemias, if that is at all possible. In any event, we have finally learned to subject all appeals for problem-solving to the preliminary queries: "Who wants the problems solved?" and "Who wants them left alone?"

The next category is perhaps the most delicate one to deal with because it falls closest to the purpose of this entire book. Theorizing about communities is a most hazardous task, as anyone who has read this far has probably discovered. A great many choices have to be made and the intellectual as well as the personal proclivities of the theorist remain all too nakedly exposed. One currently fashionable conception of sociological theory, embodied most fruitfully in the work of Talcott Parsons, and usually parodied unwittingly by his students, advocates the search for a general theory that will cover all empirical social systems. There is too much substantive material lying around the workshop of the community

[333]

sociologist to permit sustained dedication to this ideal unless the theorist is willing to sacrifice the most interesting contributions made by the field workers upon whom he depends.

For the time being the community sociologist who wants to establish continuities with earlier studies and to retain vital contact with the problems and people in contemporary communities, will have to be satisfied with simpler and more tentative theoretical models. The theorizing in this book has deliberately remained firmly attached to observed patterns of social change, sacrificing both high-level generalization and logical systematization whenever necessary to achieve historically relevant interpretations of specific case materials. The resulting theories will never be as orderly or as portable as the set of logically interrelated propositions that systematic sociologists of all ranges seek, but they seem to allow a lot more opportunity for interesting discoveries. One must be willing to forego conceptual certainties if one wishes to enjoy research adventures and human encounters.

The final role for the community sociologist is the most precarious of all. One can hardly appoint oneself a social philosopher; indeed, it is important to be wary of those who try to do so. Community sociologists are extremely fortunate in having an authenticated social philosopher in the person of Lewis Mumford, who has addressed himself in an original fashion to most of the significant problems in which we might be interested. His great series, *Technics and Civilization, The Culture of Cities, The Condition of Man,* and *The Conduct of Life,* and his numerous books of essays contain enough philosophical, historical, and sociological material to keep a team of specialists occupied for many years codifying them.

One explanation for our failure to use the wealth of material that Mumford provides probably lies in the excessive emphasis on systematic work within the profession. This is most unfortunate but there are signs that we are approaching a point where his kind of contribution can begin to be appreciated.

He employs a consistently dramatic approach to communities, viewing them as settings in which social performances are staged. His grasp of architectural styles and their symbolic meanings makes this singularly easy, and he discerns relationships between the art, architecture, technology, economic forms, political forms, religious forms, and philosophies of various historical periods with keen sociological insight. He also offers a serviceable typology of city forms and a devastating critique of modern community systems followed by a serious effort at utopian visions of possible community forms. It is all there and has been available for quite a while, waiting for us to grow up to it.

Other contemporary social philosophers whose work has important bearing on the theory of communities include Hannah Arendt, Martin Buber, Karl Jaspers, and Erich Kahler.[8] My own efforts at formulating criteria for assessing the quality of community life depended a great deal upon the theories of Sigmund Freud and Paul Radin as well as upon the work of the above-mentioned social philosophers. Following established sociological practice, my philosophical position was smuggled into Part III, loosely disguised as chapters on anthropology and psychoanalysis.

By now the fundamentals of this position ought to be clear. It rests upon the assumption that human communities exist to provide their members with full opportunities for personal development through social experimentation. This experimentation presupposes sufficient openness in personal identity so that an expanding range of possibilities is appreciated, sufficient closure in personal identity so that an integral personal

---

[8] Students of sociology who are searching for contemporary theorists who deal with the issues raised by classical social theory would do well to examine the work of these social philosophers. Such books as Arendt's *The Human Condition*, Buber's *Paths in Utopia*, Jaspers' *Man in the Modern Age*, and Kahler's *Man the Measure*, are far closer to the central concerns of Marx, Weber, Durkheim, Mannheim, and Simmel than most of the material presented in panels on social theory at the meetings of the American Sociological Society.

style gradually evolves, and sufficient dramatic perspective so that alien styles espoused by others can be appreciated without weakening one's own commitments. This conception falls close to Erikson's views on ego identity and ego integrity as well as to Sullivan's emphasis on a full repertory of interpersonal relations. It stresses the communal settings in which these goals are realized or frustrated more than either psychiatric formulation but affirms the values expressed in both of them.

The alert reader will see resemblances between this formulation and the assertion that several styles of community sociology must be cultivated. Sociology still needs a dramatist of Sigmund Freud's stature who can identify and interpret the manifold dramas of modern social life in terms comparable to Freud's analysis of the Oedipus complex. Many people once hoped that Karl Marx's brilliant analysis of capitalism in terms of commodity fetishism and class conflict would serve this purpose. Even with Max Weber's heroic modifications, the script still lacks Freud's depth, accuracy, and fruitfulness. For the time being we have to be satisfied with partial visions, but these are certainly better than no vision at all.

In retrospect, the situation of the community sociologist appears to be quite promising. The processes of urbanization, industrialization, and bureaucratization that shape modern communities have been studied and type-cases identified. It is possible now to explore these same processes in other national contexts, using American type-cases from different periods as models for comparative analysis. This should yield new insight into the processes themselves as well as into the communities in which they are studied. More research is needed into the present-day urban constellation in America. We need to identify and explore the emerging sub-communities and, particularly, to search out sources of innovation. Our research equipment is considerably more elaborate than any possessed by the early community sociologists. This asset could turn into a liability if preoccupation with technical requirements shifts attention from the fundamental issues involved in learning to understand the meanings and trace the objective structures that communities display. In light of the threatening character of alien meanings, the tendency to employ elaborate technology or elaborate theories as defenses against social insight can never be ignored. But here, too, we are learning to use the subjective responses of the observer as an instrument for improving the quality and objectivity of fieldwork. The development of a dramaturgic frame of reference promises to facilitate research and theorizing in this area as well as in the rest of sociology.

From this standpoint, the future of community studies does appear bright. However, the real fate of this subdivision of sociology, as well as of other subdivisions, depends finally upon the state of the sociological imagination. Sociology must attract, encourage, and train people who enjoy interpreting historical processes. In the final analysis, this subdivision, like any other, must be judged by its contribution to our growing knowledge about historical processes and problems. An imaginative community sociologist can study these processes and enjoy the multiple satisfactions of being an ethnographer, a problem-solver, a theorist, and a social philosopher while pursuing his craft. This is simply another way of saying that he can still be a *human scientist*.

BIBLIOGRAPHY (Selective)

Anderson, Elin E., *We Americans*, Cambridge, Mass.: Harvard University Press, 1937.

Benedict, Ruth, "Continuities and Discontinuities in Cultural Conditioning" in *A Study of Interpersonal Relations*, edited by Patrick Mullahy, New York, N.Y.: Hermitage Press, Inc., 1949, pp. 297-308.

Davis, Allison, Gardner, Burleigh B., and Gardner, Mary R., *Deep South*, Chicago, Ill.: The University of Chicago Press, 1941.

Dollard, John, *Caste and Class in a Southern Town*, Garden City, N.Y.: Doubleday Anchor Books, 1957.

Durkheim, Emile, *The Division of Labor in Society*, Glencoe, Ill.: The Free Press, 1947.

Durkheim, Emile, *Suicide*, Glencoe, Ill.: The Free Press, 1951.

Erikson, Erik Homburger, *Childhood and Society*, New York, N.Y.: W. W. Norton and Co., Inc., 1950.

Erikson, Erik Homburger, "The Problem of Ego Identity," *Journal of the American Psychoanalytic Association*, Vol. IV, 1956, pp. 56-121. Also in "Identity and Anxiety," ed. by Stein, Vidich, and White.

Evans-Pritchard, E. E., et al., *The Institutions of Primitive Society*, Glencoe, Ill.: The Free Press, 1956.

Faris, Robert and Dunham, H. Warren, *Mental Disorders in Urban Areas*, Chicago, Ill.: The University of Chicago Press, 1939.

Freud, Sigmund, *A General Introduction to Psychoanalysis*, Garden City, N.Y.: Garden City Publishing Co., Inc., 1943.

Fromm, Erich, *Man for Himself*, New York, N.Y.: Rinehart and Co., Inc., 1947.

Goffman, Erving, *The Presentation of Self in Everyday Life*, Garden City, N.Y.: Doubleday Anchor Books, 1959.

Goodman, Percival, and Goodman, Paul, *Communitas*, Chicago, Ill.: The University of Chicago Press, 1947.

Gouldner, Alvin W., *Patterns of Industrial Bureaucracy*, Glencoe, Ill.: The Free Press, 1954.

Hatt, Paul K., and Reiss, Albert J., Jr., *Reader in Urban Society*, Glencoe, Ill.: The Free Press, 1951.

Hughes, Everett Cherrington, *Men and Their Work*, Glencoe, Ill.: The Free Press, 1958.

Hughes, H. Stuart, *Consciousness and Society*, New York, N.Y.: Knopf, 1958.

Hunter, Floyd, *Community Power Structure*, Chapel Hill, N.C.: The University of North Carolina Press, 1953.

Jaspers, Karl, *Man in the Modern Age*, London: Routledge and Kegan Paul Ltd., new edition, 1951.

Lewin, Kurt, *Field Theory in Social Science*, New York, N.Y.: Harper and Brothers, 1951.

Lynd, Robert, and Lynd, Helen, *Middletown*, New York, N.Y.: Harcourt-Brace and Co., 1929.

Lynd, Robert, and Lynd, Helen, *Middletown in Transition*, New York, N.Y.: Harcourt-Brace and Co., 1937.

Mannheim, Karl, *Essays on Sociology and Social Psychology*, New York, N.Y.: Oxford University Press, 1953.

Mauldin, Bill, *Up Front*, New York, N.Y.: The World Publishing Co., 1945.

Merton, Robert K., *Social Theory and Social Structure*, Glencoe, Ill.: The Free Press, Revised and Enlarged Edition, 1957.

Mills, C. Wright, *The Power Elite*, New York, N.Y.: Oxford University Press, 1956.

Mills, C. Wright, *The Sociological Imagination*, New York, N.Y.: Oxford University Press, 1959.

Moore, Wilbert, *Industrialization and Labor*, Ithaca, N.Y.: Cornell University Press, 1951.

Mumford, Lewis, *The Culture of Cities*, New York, N.Y.: Harcourt-Brace and Co., 1938.

Park, Robert E., *Human Communities*, Glencoe, Ill.: The Free Press, 1952.

Park, Robert E., *Race and Culture*, Glencoe, Ill.: The Free Press, 1950.

Park, Robert E., *Society*, Glencoe, Ill.: The Free Press, 1954.

Radin, Paul, *Primitive Man as Philosopher*, New York, N.Y.: Appleton, 1927.

Radin, Paul, *The World of Primitive Man*, New York, N.Y.: Henry Schuman, 1953.

Redfield, Robert, *The Little Community*, Chicago, Ill.: The University of Chicago Press, 1955.

Redfield, Robert, *Peasant Society and Culture*, Chicago, Ill.: The University of Chicago Press, 1956.

Riesman, David, *The Lonely Crowd*, New Haven, Conn.: Yale University Press, 1950.

Sapir, Edward, *Selected Writings of Edward Sapir*, edited by David G. Mandelbaum, Berkeley, Calif.: University of California Press, 1949.

Schachtel, Ernest, "On Memory and Childhood Amnesia" in *A Study of Interpersonal Relations*, edited by Patrick Mullahy, New York, N.Y.: Hermitage Press, Inc., 1949, pp. 3-49.

Seeley, John R., Sim, R. Alexander, and Loosley, Elizabeth W., *Crestwood Heights*, New York, N.Y.: Basic Books, Inc., 1956.

Shaw, Clifford R., *The Jackroller*, Chicago, Ill.: The University of Chicago Press, 1930.

Shaw, Clifford R., et al., *Delinquency Areas*, Chicago, Ill.: The University of Chicago Press, 1929.

Simmel, Georg, *The Sociology of Georg Simmel*, Glencoe, Ill.: The Free Press, 1950. Translated by Kurt Wolff.

Spectorsky, A. C., *The Exurbanites*, New York, N.Y.: Berkley Publishing Corp., 1958 (paperback).

Speier, Hans, *Social Order and the Risks of War*, New York, N.Y.: George W. Stewart, Publisher, Inc., 1952.

Stein, Maurice R., Vidich, Arthur, and White, David, *Identity and Anxiety*, Glencoe, Ill.: The Free Press, 1960.

Stonequist, E. V., *The Marginal Man*, New York, N.Y.: Charles Scribner's Sons, 1937.

Stouffer, Samuel A., et al., *The American Soldier*, Volumes ɪ and ɪɪ, Princeton, N.J.: Princeton University Press, 1949.

Sullivan, Harry Stack, *The Interpersonal Theory of Psychiatry*, New York, N.Y.: W. W. Norton and Co., 1953.

Thrasher, Frederick, *The Gang*, Chicago, Ill.: The University of Chicago Press, 1927.

Veblen, Thorstein, *The Theory of the Leisure Class*, New York, N.Y.: Random House, 1934.

Vidich, Arthur J., and Bensman, Joseph, *Small Town in Mass Society*, Princeton, N.J.: Princeton University Press, 1958.

Ware, Caroline C., *Greenwich Village*, New York, N.Y.: Houghton Mifflin Co., 1935.

Warner, W. Lloyd, and Lunt, Paul S., *The Social Life of a Modern Community*, New Haven, Conn.: Yale University Press, 1941.

Warner, W. Lloyd, and Srole, Leo, *The Social System of American Ethnic Groups*, New Haven, Conn.: Yale University Press, 1945.

Warner, W. Lloyd, and Low, J. O., *The Social System of the Modern Factory*, New Haven, Conn.: Yale University Press, 1947.

Watson, David Lindsay, *The Study of Human Nature*, Yellow Springs, Ohio: The Antioch Press, 1953.

Weber, Max, *Essays in Sociology*, New York, N.Y.: Oxford University Press, 1946. Translated by Hans Gerth and C. Wright Mills.

Whyte, William Foote, *Street Corner Society*, Chicago, Ill.: The University of Chicago Press, 2nd edition, 1955.

Whyte, William H., *The Organization Man*, Garden City, N.Y.: Doubleday Anchor Books, 1957.

Wirth, Louis, *The Ghetto*, Chicago, Ill.: The University of Chicago Press, 1928.

Zorbaugh, Harvey, *The Gold Coast and the Slum*, Chicago, Ill.: The University of Chicago Press, 1929.

NAME INDEX

SUBJECT INDEX

70 71 72 73   12 11 10 9 8 7 6

Revised December, 1967

harper ✲ torchbooks

HUMANITIES AND SOCIAL SCIENCES

American Studies: General

LOUIS D. BRANDEIS: Other People's Money, and How the Bankers Use It. ‡ Ed. with an Intro. by Richard M. Abrams
TB/3081

THOMAS C. COCHRAN; The Inner Revolution. Essays on the Social Sciences in History
TB/1140

HENRY STEELE COMMAGER, Ed.: The Struggle for Racial Equality
TB/1300

EDWARD S. CORWIN: American Constitutional History. Essays edited by Alpheus T. Mason and Gerald Garvey △
TB/1136

CARL N. DEGLER, Ed.: Pivotal Interpretations of American History   Vol. I TB/1240; Vol. II TB/1241

A. HUNTER DUPREE: Science in the Federal Government: A History of Policies and Activities to 1940   TB/573

A. S. EISENSTADT, Ed.: The Craft of American History: Recent Essays in American Historical Writing
Vol. I TB/1255; Vol. II TB/1256

CHARLOTTE P. GILMAN: Women and Economics: A Study of the Economic Relation between Men and Women as a Factor in Social Evolution. ‡ Ed. with an Introduction by Carl N. Degler
TB/3073

OSCAR HANDLIN: This Was America: As Recorded by European Travelers in the Eighteenth, Nineteenth and Twentieth Centuries. Illus.
TB/1119

MARCUS LEE HANSEN: The Atlantic Migration: 1607-1860. Edited by Arthur M. Schlesinger
TB/1052

MARCUS LEE HANSEN: The Immigrant in American History.
TB/1120

JOHN HIGHAM, Ed.: The Reconstruction of American History △
TB/1068

ROBERT H. JACKSON: The Supreme Court in the American System of Government
TB/1106

JOHN F. KENNEDY: A Nation of Immigrants. △ Illus.
TB/1118

LEONARD W. LEVY, Ed.: American Constitutional Law: Historical Essays
TB/1285

LEONARD W. LEVY, Ed.: Judicial Review and the Supreme Court
TB/1296

LEONARD W. LEVY: The Law of the Commonwealth and Chief Justice Shaw
TB/1309

HENRY F. MAY: Protestant Churches and Industrial America. New Intro. by the Author
TB/1334

RALPH BARTON PERRY: Puritanism and Democracy
TB/1138

ARNOLD ROSE: The Negro in America
TB/3048

MAURICE R. STEIN: The Eclipse of Community. An Interpretation of American Studies
TB/1128

W. LLOYD WARNER and Associates: Democracy in Jonesville: A Study in Quality and Inequality ¶   TB/1129

W. LLOYD WARNER: Social Class in America: The Evaluation of Status
TB/1013

American Studies: Colonial

BERNARD BAILYN, Ed.: Apologia of Robert Keayne: Self-Portrait of a Puritan Merchant
TB/1201

BERNARD BAILYN: The New England Merchants in the Seventeenth Century
TB/1149

JOSEPH CHARLES: The Origins of the American Party System
TB/1049

HENRY STEELE COMMAGER & ELMO GIORDANETTI, Eds.: Was America a Mistake? An Eighteenth Century Controversy
TB/1329

CHARLES GIBSON: Spain in America †
TB/3077

LAWRENCE HENRY GIPSON: The Coming of the Revolution: 1763-1775. † Illus.
TB/3007

LEONARD W. LEVY: Freedom of Speech and Press in Early American History: Legacy of Suppression   TB/1109

PERRY MILLER: Errand Into the Wilderness
TB/1139

PERRY MILLER & T. H. JOHNSON, Eds.: The Puritans: A Sourcebook of Their Writings
Vol. I TB/1093; Vol. II TB/1094

EDMUND S. MORGAN, Ed.: The Diary of Michael Wigglesworth, 1653-1657: The Conscience of a Puritan
TB/1228

EDMUND S. MORGAN: The Puritan Family: Religion and Domestic Relations in Seventeenth-Century New England
TB/1227

RICHARD B. MORRIS: Government and Labor in Early America
TB/1244

KENNETH B. MURDOCK: Literature and Theology in Colonial New England
TB/99

WALLACE NOTESTEIN: The English People on the Eve of Colonization: 1603-1630. † Illus.
TB/3006

JOHN P. ROCHE: Origins of American Political Thought: Selected Readings
TB/1301

JOHN SMITH: Captain John Smith's America: Selections from His Writings. Ed. with Intro. by John Lankford
TB/3078

LOUIS B. WRIGHT: The Cultural Life of the American Colonies: 1607-1763. † Illus.
TB/3005

American Studies: From the Revolution to 1860

JOHN R. ALDEN: The American Revolution: 1775-1783. † Illus.
TB/3011

MAX BELOFF, Ed.: The Debate on the American Revolution, 1761-1783: A Sourcebook △
TB/1225

RAY A. BILLINGTON: The Far Western Frontier: 1830-1860. † Illus.
TB/3012

EDMUND BURKE: On the American Revolution: Selected Speeches and Letters. ‡ Edited by Elliott Robert Barkan
TB/3068

WHITNEY R. CROSS: The Burned-Over District: The Social and Intellectual History of Enthusiastic Religion in Western New York, 1800-1850 △
TB/1242

GEORGE DANGERFIELD: The Awakening of American Nationalism: 1815-1828. † Illus.
TB/3061

---

† The New American Nation Series, edited by Henry Steele Commager and Richard B. Morris.
‡ American Perspectives series, edited by Bernard Wishy and William E. Leuchtenburg.
* The Rise of Modern Europe series, edited by William L. Langer.
** History of Europe series, edited by J. H. Plumb.
¶ Researches in the Social, Cultural and Behavioral Sciences, edited by Benjamin Nelson.
§ The Library of Religion and Culture, edited by Benjamin Nelson.
Σ Harper Modern Science Series, edited by James R. Newman.
° Not for sale in Canada.
△ Not for sale in the U. K.

3

G. G. COULTON: Medieval Village, Manor, and Monastery
TB/1022
CHRISTOPHER DAWSON, Ed.: Mission to Asia: Narratives and Letters of the Franciscan Missionaries in Mongolia and China in the 13th and 14th Centuries △
TB/315
HEINRICH FICHTENAU: The Carolingian Empire: The Age of Charlemagne △
TB/1142
GALBERT OF BRUGES: The Murder of Charles the Good. Trans. with Intro. by James Bruce Ross
TB/1311
F. L. GANSHOF: Feudalism △
TB/1058
DENO GEANAKOPLOS: Byzantine East and Latin West: Two Worlds of Christendom in the Middle Ages and Renaissance
TB/1265
EDWARD GIBBON: The Triumph of Christendom in the Roman Empire (Chaps. XV-XX of "Decline and Fall," J. B. Bury edition). § △ Illus.
TB/46
W. O. HASSALL, Ed.: Medieval England: As Viewed by Contemporaries △
TB/1205
DENYS HAY: Europe: The Emergence of an Idea TB/1275
DENYS HAY: The Medieval Centuries ° △
TB/1192
J. M. HUSSEY: The Byzantine World △
TB/1057
ROBERT LATOUCHE: The Birth of Western Economy: Economic Aspects of the Dark Ages. ° △ Intro. by Philip Grierson
TB/1290
FERDINAND LOT: The End of the Ancient World and the Beginnings of the Middle Ages. Introduction by Glanville Downey
TB/1044
ACHILLE LUCHAIRE: Social France at the Time of Philip Augustus. New Intro. by John W. Baldwin   TB/1314
MARSILIUS OF PADUA: The Defender of the Peace. Trans. with Intro. by Alan Gewirth
TB/1310
G. MOLLAT: The Popes at Avignon: 1305-1378 △ TB/308
CHARLES PETIT-DUTAILLIS: The Feudal Monarchy in France and England: From the Tenth to the Thirteenth Century ° △
TB/1165
HENRI PIRENNE: Early Democracies in the Low Countries: Urban Society and Political Conflict in the Middle Ages and the Renaissance. Introduction by John H. Mundy
TB/1110
STEVEN RUNCIMAN: A History of the Crusades. △
Volume I: The First Crusade and the Foundation of the Kingdom of Jerusalem. Illus.
TB/1143
Volume II: The Kingdom of Jerusalem and the Frankish East, 1100-1187. Illus.
TB/1243
Volume III: The Kingdom of Acre and the Later Crusades
TB/1298
SULPICIUS SEVERUS et al.: The Western Fathers: Being the Lives of Martin of Tours, Ambrose, Augustine of Hippo, Honoratus of Arles and Germanus of Auxerre. △ Edited and trans. by F. O. Hoare   TB/309
J. M. WALLACE-HADRILL: The Barbarian West: The Early Middle Ages, A.D. 400-1000 △
TB/1061

History: Renaissance & Reformation

JACOB BURCKHARDT: The Civilization of the Renaissance in Italy. △ Intro. by Benjamin Nelson & Charles Trinkaus. Illus.   Vol. I TB/40; Vol. II TB/41
JOHN CALVIN & JACOPO SADOLETO: A Reformation Debate. Edited by John C. Olin
TB/1239
ERNST CASSIRER: The Individual and the Cosmos in Renaissance Philosophy. △ Translated with an Introduction by Mario Domandi
TB/1097
FEDERICO CHABOD: Machiavelli and the Renaissance △
TB/1193
EDWARD P. CHEYNEY: The Dawn of a New Era, 1250-1453. * Illus.
TB/3002
G. CONSTANT: The Reformation in England: The English Schism, Henry VIII, 1509-1547 △
TB/314
R. TREVOR DAVIES: The Golden Century of Spain, 1501-1621 ° △
TB/1194
G. R. ELTON: Reformation Europe, 1517-1559 ** ° △
TB/1270

DESIDERIUS ERASMUS: Christian Humanism and the Reformation: Selected Writings. Edited and translated by John C. Olin
TB/1166
WALLACE K. FERGUSON et al.: Facets of the Renaissance
TB/1098
WALLACE K. FERGUSON et al.: The Renaissance: Six Essays. Illus.
TB/1084
JOHN NEVILLE FIGGIS: The Divine Right of Kings. Introduction by G. R. Elton
TB/1191
JOHN NEVILLE FIGGIS: Political Thought from Gerson to Grotius: 1414-1625: Seven Studies. Introduction by Garrett Mattingly
TB/1032
MYRON P. GILMORE: The World of Humanism, 1453-1517. * Illus.
TB/3003
FRANCESCO GUICCIARDINI: Maxims and Reflections of a Renaissance Statesman (Ricordi). Trans. by Mario Domandi. Intro. by Nicolai Rubinstein   TB/1160
J. H. HEXTER: More's Utopia: The Biography of an Idea. New Epilogue by the Author
TB/1195
HAJO HOLBORN: Ulrich von Hutten and the German Reformation
TB/1238
JOHAN HUIZINGA: Erasmus and the Age of Reformation. △ Illus.
TB/19
JOEL HURSTFIELD: The Elizabethan Nation △   TB/1312
JOEL HURSTFIELD, Ed.: The Reformation Crisis △ TB/1267
ULRICH VON HUTTEN et al.: On the Eve of the Reformation: "Letters of Obscure Men." Introduction by Hajo Holborn
TB/1124
PAUL O. KRISTELLER: Renaissance Thought: The Classic, Scholastic, and Humanist Strains
TB/1048
PAUL O. KRISTELLER: Renaissance Thought II: Papers on Humanism and the Arts
TB/1163
NICCOLÒ MACHIAVELLI: History of Florence and of the Affairs of Italy: from the earliest times to the death of Lorenzo the Magnificent. △ Introduction by Felix Gilbert
TB/1027
ALFRED VON MARTIN: Sociology of the Renaissance. Introduction by Wallace K. Ferguson
TB/1099
GARRETT MATTINGLY et al.: Renaissance Profiles. △ Edited by J. H. Plumb
TB/1162
MILLARD MEISS: Painting in Florence and Siena after the Black Death: The Arts, Religion and Society in the Mid-Fourteenth Century. △ 169 illus.
TB/1148
J. E. NEALE: The Age of Catherine de Medici ° △ TB/1085
ERWIN PANOFSKY: Studies in Iconology: Humanistic Themes in the Art of the Renaissance. △ 180 illustrations
TB/1077
J. H. PARRY: The Establishment of the European Hegemony: 1415-1715: Trade and Exploration in the Age of the Renaissance △
TB/1045
BUONACCORSO PITTI & GREGORIO DATI: Two Memoirs of Renaissance Florence: The Diaries of Buonaccorso Pitti and Gregorio Dati. Ed. with an Intro. by Gene Brucker. Trans. by Julia Martines
TB/1333
J. H. PLUMB: The Italian Renaissance: A Concise Survey of Its History and Culture △
TB/1161
A. F. POLLARD: Henry VIII. ° △ Introduction by A. G. Dickens
TB/1249
A. F. POLLARD: Wolsey. ° △ Introduction by A. G. Dickens
TB/1248
CECIL ROTH: The Jews in the Renaissance. Illus.   TB/834
A. L. ROWSE: The Expansion of Elizabethan England. ° △ Illus.
TB/1220
GORDON RUPP: Luther's Progress to the Diet of Worms ° △
TB/120
FERDINAND SCHEVILL: The Medici. Illus.
TB/1010
FERDINAND SCHEVILL: Medieval and Renaissance Florence. Illus.   Volume I: Medieval Florence TB/1090
Volume II: The Coming of Humanism and the Age of the Medici
TB/1091
R. H. TAWNEY: The Agrarian Problem in the Sixteenth Century. New Intro. by Lawrence Stone   TB/1315
G. M. TREVELYAN: England in the Age of Wycliffe, 1368-1520 ° △
TB/1112

4

'VESPASIANO: Renaissance Princes, Popes, and Prelates: *The Vespasiano Memoirs: Lives of Illustrious Men of the XVth Century. Intro. by Myron P. Gilmore*
TB/1111

History: Modern European

FREDERICK B. ARTZ: Reaction and Revolution, 1815-1832. * Illus. TB/3034
MAX BELOFF: The Age of Absolutism, 1660-1815 △
TB/1062
ROBERT C. BINKLEY: Realism and Nationalism, 1852-1871. * Illus. TB/3038
EUGENE C. BLACK, Ed.: European Political History, 1815-1870: *Aspects of Liberalism* TB/1331
ASA BRIGGS: The Making of Modern England, 1784-1867: *The Age of Improvement* ° △ TB/1203
CRANE BRINTON: A Decade of Revolution, 1789-1799. * Illus. TB/3018
D. W. BROGAN: The Development of Modern France. ° △ Volume I: *From the Fall of the Empire to the Dreyfus Affair* TB/1184
Volume II: *The Shadow of War, World War I, Between the Two Wars. New Introduction by the Author* TB/1185
J. BRONOWSKI & BRUCE MAZLISH: The Western Intellectual Tradition: *From Leonardo to Hegel* △ TB/3001
GEOFFREY BRUUN: Europe and the French Imperium, 1799-1814. * Illus. TB/3033
ALAN BULLOCK: Hitler, A Study in Tyranny. ° △ *Illus.* TB/1123
E. H. CARR: German-Soviet Relations Between the Two World Wars, 1919-1939 TB/1278
E. H. CARR: International Relations Between the Two World Wars, 1919-1939 ° △ TB/1279
E. H. CARR: The Twenty Years' Crisis, 1919-1939: *An Introduction to the Study of International Relations* ° △ TB/1122
GORDON A. CRAIG: From Bismarck to Adenauer: *Aspects of German Statecraft. Revised Edition* TB/1171
DENIS DIDEROT: The Encyclopedia: *Selections. Ed. and trans. by Stephen Gendzier* TB/1299
WALTER L. DORN: Competition for Empire, 1740-1763. * Illus. TB/3032
FRANKLIN L. FORD: Robe and Sword: *The Regrouping of the French Aristocracy after Louis XIV* TB/1217
CARL J. FRIEDRICH: The Age of the Baroque, 1610-1660. * Illus. TB/3004
RENÉ FUELOEP-MILLER: The Mind and Face of Bolshevism: *An Examination of Cultural Life in Soviet Russia. New Epilogue by the Author* TB/1188
M. DOROTHY GEORGE: London Life in the Eighteenth Century △ TB/1182
LEO GERSHOY: From Despotism to Revolution, 1763-1789. * Illus. TB/3017
C. C. GILLISPIE: Genesis and Geology: *The Decades before Darwin* § TB/51
ALBERT GOODWIN, Ed.: The European Nobility in the Eighteenth Century △ TB/1313
ALBERT GOODWIN: The French Revolution △ TB/1064
ALBERT GUÉRARD: France in the Classical Age: *The Life and Death of an Ideal* △ TB/1183
CARLTON J. H. HAYES: A Generation of Materialism, 1871-1900. * Illus. TB/3039
J. H. HEXTER: Reappraisals in History: *New Views on History and Society in Early Modern Europe* △ TB/1100
STANLEY HOFFMANN et al.: In Search of France: *The Economy, Society and Political System in the Twentieth Century* TB/1219
A. R. HUMPHREYS: The Augustan World: *Society, Thought, & Letters in 18th Century England* ° △ TB/1105
DAN N. JACOBS, Ed.: The New Communist Manifesto and Related Documents. Third edition, revised TB/1078

LIONEL KOCHAN: The Struggle for Germany: *1914-45* TB/1304
HANS KOHN: The Mind of Germany: *The Education of a Nation* △ TB/1204
HANS KOHN, Ed.: The Mind of Modern Russia: *Historical and Political Thought of Russia's Great Age* TB/1065
WALTER LAQUEUR & GEORGE L. MOSSE, Eds.: Education and Social Structure in the 20th Century. ° △ *Vol. 6 of the* Journal of Contemporary History TB/1339
WALTER LAQUEUR & GEORGE L. MOSSE, Eds.: International Fascism, 1920-1945. ° △ *Volume 1 of* Journal of Contemporary History TB/1276
WALTER LAQUEUR & GEORGE L. MOSSE, Eds.: The Left-Wing Intellectuals between the Wars 1919-1939. ° △ *Volume 2 of* Journal of Contemporary History TB/1286
WALTER LAQUEUR & GEORGE L. MOSSE, Eds.: Literature and Politics in the 20th Century. ° △ *Vol. 5 of the* Journal of Contemporary History TB/1328
WALTER LAQUEUR & GEORGE L. MOSSE, Eds.: The New History: *Trends in Historical Research and Writing since World War II.* ° △ *Vol. 4 of the* Journal of Contemporary History TB/1327
WALTER LAQUEUR & GEORGE L. MOSSE, Eds.: 1914: *The Coming of the First World War.* ° △ *Volume 3 of* Journal of Contemporary History TB/1306
FRANK E. MANUEL: The Prophets of Paris: *Turgot, Condorcet, Saint-Simon, Fourier, and Comte* TB/1218
KINGSLEY MARTIN: French Liberal Thought in the Eighteenth Century: *A Study of Political Ideas from Bayle to Condorcet* TB/1114
ROBERT K. MERTON: Science, Technology and Society in Seventeenth Century England ¶ *New Intro. by the Author* TB/1324
L. B. NAMIER: Facing East: *Essays on Germany, the Balkans, and Russia in the 20th Century* △ TB/1280
L. B. NAMIER: Personalities and Powers: *Selected Essays* △ TB/1186
L. B. NAMIER: Vanished Supremacies: *Essays on European History, 1812-1918* ° TB/1088
NAPOLEON III: Napoleonic Ideas: *Des Idées Napoléoniennes, par le Prince Napoléon-Louis Bonaparte. Ed. by Brison D. Gooch* TB/1336
FRANZ NEUMANN: Behemoth: *The Structure and Practice of National Socialism, 1933-1944* TB/1289
FREDERICK L. NUSSBAUM: The Triumph of Science and Reason, 1660-1685. * Illus. TB/3009
DAVID OGG: Europe of the Ancien Régime, 1715-1783 ** ° △ TB/1271
JOHN PLAMENATZ: German Marxism and Russian Communism. ° *New Preface by the Author* TB/1189
RAYMOND W. POSTGATE, Ed.: Revolution from 1789 to 1906: *Selected Documents* TB/1063
PENFIELD ROBERTS: The Quest for Security, 1715-1740. * Illus. TB/3016
PRISCILLA ROBERTSON: Revolutions of 1848: *A Social History* TB/1025
GEORGE RUDÉ: Revolutionary Europe, 1783-1815 ** ° △ TB/1272
LOUIS, DUC DE SAINT-SIMON: Versailles, The Court, and Louis XIV. ° △ *Introductory Note by Peter Gay* TB/1250
HUGH SETON-WATSON: Eastern Europe Between the Wars, 1918-1941 TB/1330
ALBERT SOREL: Europe Under the Old Regime. *Translated by Francis H. Herrick* TB/1121
N. N. SUKHANOV: The Russian Revolution, 1917: *Eyewitness Account.* △ *Edited by Joel Carmichael* Vol. I TB/1066; Vol. II TB/1067
A. J. P. TAYLOR: From Napoleon to Lenin: *Historical Essays* ° △ TB/1268
A. J. P. TAYLOR: The Habsburg Monarchy, 1809-1918: *A History of the Austrian Empire and Austria-Hungary* ° △ TB/1187
G. M. TREVELYAN: British History in the Nineteenth Century and After: 1782-1919. ° △ *Second Edition* TB/1251

5

7

HANS KOHN: Political Ideologies of the 20th Century
TB/1277
ROY C. MACRIDIS, Ed.: Political Parties: *Contemporary Trends and Ideas* TB/1322
ROBERT GREEN MC CLOSKEY: American Conservatism in the Age of Enterprise, 1865-1910 TB/1137
KINGSLEY MARTIN: French Liberal Thought in the Eighteenth Century: *Political Ideas from Bayle to Condorcet* △ TB/1114
ROBERTO MICHELS: First Lectures in Political Sociology. *Edited by Alfred de Grazia* ¶ ° TB/1224
JOHN STUART MILL: On Bentham and Coleridge. △ *Introduction by F. R. Leavis* TB/1070
BARRINGTON MOORE, JR.: Political Power and Social Theory: *Seven Studies* ¶ TB/1221
BARRINGTON MOORE, JR.: Soviet Politics—The Dilemma of Power: *The Role of Ideas in Social Change* ¶
TB/1222
BARRINGTON MOORE, JR.: Terror and Progress—USSR: *Some Sources of Change and Stability in the Soviet Dictatorship* ¶ TB/1266
JOHN B. MORRALL: Political Thought in Medieval Times △ TB/1076
JOHN PLAMENATZ: German Marxism and Russian Communism. ° △ *New Preface by the Author* TB/1189
KARL R. POPPER: The Open Society and Its Enemies △
Vol. I: *The Spell of Plato* TB/1101
Vol. II: *The High Tide of Prophecy: Hegel, Marx and the Aftermath* TB/1102
JOHN P. ROCHE, Ed.: American Political Thought: *From Jefferson to Progressivism* TB/1332
HENRI DE SAINT-SIMON: Social Organization, The Science of Man, and Other Writings. *Edited and Translated by Felix Markham* TB/1152
CHARLES I. SCHOTTLAND, Ed.: The Welfare State TB/1323
JOSEPH A. SCHUMPETER: Capitalism, Socialism and Democracy △ TB/3008
BENJAMIN I. SCHWARTZ: Chinese Communism and the Rise of Mao TB/1308
CHARLES H. SHINN: Mining Camps: *A Study in American Frontier Government.* ‡ *Edited by Rodman W. Paul*
TB/3062
PETER WOLL, Ed.: Public Administration and Policy: *Selected Essays* TB/1284

Psychology

ALFRED ADLER: The Individual Psychology of Alfred Adler. △ *Edited by Heinz L. and Rowena R. Ansbacher*
TB/1154
ALFRED ADLER: Problems of Neurosis. *Introduction by Heinz L. Ansbacher* TB/1145
ARTHUR BURTON & ROBERT E. HARRIS, Eds.: Clinical Studies of Personality
Vol. I TB/3075; Vol. II TB/3076
HADLEY CANTRIL: The Invasion from Mars: *A Study in the Psychology of Panic* ¶ TB/1282
HERBERT FINGARETTE: The Self in Transformation: *Psychoanalysis, Philosophy and the Life of the Spirit* ¶
TB/1177
SIGMUND FREUD: On Creativity and the Unconscious: *Papers on the Psychology of Art, Literature, Love, Religion.* § △ *Intro. by Benjamin Nelson* TB/45
C. JUDSON HERRICK: The Evolution of Human Nature
TB/545
WILLIAM JAMES: Psychology: *The Briefer Course. Edited with an Intro. by Gordon Allport* TB/1034
C. G. JUNG: Psychological Reflections △ TB/2001
C. G. JUNG: Symbols of Transformation: *An Analysis of the Prelude to a Case of Schizophrenia.* △ *Illus.*
Vol. I TB/2009; Vol. II TB/2010
C. G. JUNG & C. KERÉNYI: Essays on a Science of Mythology: *The Myths of the Divine Child and the Divine Maiden* TB/2014

KARL MENNINGER: Theory of Psychoanalytic Technique
TB/1144
ERICH NEUMANN: Amor and Psyche: *The Psychic Development of the Feminine* △ TB/2012
ERICH NEUMANN: The Archetypal World of Henry Moore. △ *107 illus.* TB/2020
ERICH NEUMANN: The Origins and History of Consciousness △ Vol. I Illus. TB/2007; Vol. II TB/2008
RALPH BARTON PERRY: The Thought and Character of William James: *Briefer Version* TB/1156
JOHN H. SCHAAR: Escape from Authority: *The Perspectives of Erich Fromm* TB/1155
MUZAFER SHERIF: The Psychology of Social Norms
TB/3072

Sociology

JACQUES BARZUN: Race: *A Study in Superstition. Revised Edition* TB/1172
BERNARD BERELSON, Ed.: The Behavioral Sciences Today
TB/1127
ABRAHAM CAHAN: The Rise of David Levinsky: *A documentary novel of social mobility in early twentieth century America. Intro. by John Higham* TB/1028
KENNETH B. CLARK: Dark Ghetto: *Dilemmas of Social Power. Foreword by Gunnar Myrdal* TB/1317
LEWIS A. COSER, Ed.: Political Sociology TB/1293
ALLISON DAVIS & JOHN DOLLARD: Children of Bondage: *The Personality Development of Negro Youth in the Urban South* ¶ TB/3049
ST. CLAIR DRAKE & HORACE R. CAYTON: Black Metropolis: *A Study of Negro Life in a Northern City.* △ *Revised and Enlarged. Intro. by Everett C. Hughes*
Vol. I TB/1086; Vol. II TB/1087
EMILE DURKHEIM et al.: Essays on Sociology and Philosophy: *With Analyses of Durkheim's Life and Work.* ¶ *Edited by Kurt H. Wolff* TB/1151
LEON FESTINGER, HENRY W. RIECKEN & STANLEY SCHACHTER: When Prophecy Fails: *A Social and Psychological Account of a Modern Group that Predicted the Destruction of the World* ¶ TB/1132
ALVIN W. GOULDNER: Wildcat Strike: *A Study in Worker-Management Relationships* ¶ ¶ TB/1176
CÉSAR GRAÑA: Modernity and Its Discontents: *French Society and the French Man of Letters in the Nineteenth Century* ¶ TB/1318
FRANCIS J. GRUND: Aristocracy in America: *Social Class in the Formative Years of the New Nation* △ TB/1001
KURT LEWIN: Field Theory in Social Science: *Selected Theoretical Papers.* ¶ △ *Edited with a Foreword by Dorwin Cartwright* TB/1135
R. M. MAC IVER: Social Causation TB/1153
ROBERT K. MERTON, LEONARD BROOM, LEONARD S. COTTRELL, JR., Editors: Sociology Today: *Problems and Prospects* ¶ Vol. I TB/1173; Vol. II TB/1174
ROBERTO MICHELS: First Lectures in Political Sociology. *Edited by Alfred de Grazia* ¶ ° TB/1224
BARRINGTON MOORE, JR.: Political Power and Social Theory: *Seven Studies* ¶ TB/1221
BARRINGTON MOORE, JR.: Soviet Politics—The Dilemma of Power: *The Role of Ideas in Social Change* ¶
TB/1222
TALCOTT PARSONS & EDWARD A. SHILS, Editors: Toward a General Theory of Action: *Theoretical Foundations for the Social Sciences* TB/1083
ARNOLD ROSE: The Negro in America: *The Condensed Version of Gunnar Myrdal's An American Dilemma*
TB/3048
GEORGE ROSEN: Madness in Society: *Chapters in the Historical Sociology of Mental Illness.* ¶ *Preface by Benjamin Nelson* TB/1337
KURT SAMUELSSON: Religion and Economic Action: *A Critique of Max Weber's The Protestant Ethic and the Spirit of Capitalism.* ¶ ° *Trans. by E. G. French. Ed. with Intro. by D. C. Coleman* TB/1131

9

11

A. G. VAN MELSEN: From Atomos to Atom: *A History of the Concept Atom* TB/517

STEPHEN TOULMIN & JUNE GOODFIELD: The Architecture of Matter: *Physics, Chemistry & Physiology of Matter, Both Animate & Inanimate, As it Evolved Since the Beginning of Science* ° △ TB/584

STEPHEN TOULMIN & JUNE GOODFIELD: The Discovery of Time ° △ TB/585

Mathematics

E. W. BETH: The Foundations of Mathematics: *A Study in the Philosophy of Science* △ TB/581

S. KÖRNER: The Philosophy of Mathematics: *An Introduction* △ TB/547

GEORGE E. OWEN: Fundamentals of Scientific Mathematics TB/569

WILLARD VAN ORMAN QUINE: Mathematical Logic TB/558

FREDERICK WAISMANN: Introduction to Mathematical Thinking. *Foreword by Karl Menger* TB/511

Philosophy of Science

R. B. BRAITHWAITE: Scientific Explanation TB/515

J. BRONOWSKI: Science and Human Values. △ *Revised and Enlarged Edition* TB/505

ALBERT EINSTEIN et al.: Albert Einstein: Philosopher-Scientist. *Edited by Paul A. Schilpp* Vol. I TB/502
Vol. II TB/503

WERNER HEISENBERG: Physics and Philosophy: *The Revolution in Modern Science* △ TB/549

KARL R. POPPER: Logic of Scientific Discovery △ TB/576

STEPHEN TOULMIN: Foresight and Understanding: *An Enquiry into the Aims of Science.* △ *Foreword by Jacques Barzun* TB/564

STEPHEN TOULMIN: The Philosophy of Science: *An Introduction* △ TB/513

Physics and Cosmology

JOHN E. ALLEN: Aerodynamics: *A Space Age Survey* △ TB/582

P. W. BRIDGMAN: Nature of Thermodynamics TB/537

C. V. DURELL: Readable Relativity. △ *Foreword by Freeman J. Dyson* TB/530

ARTHUR EDDINGTON: Space, Time and Gravitation: *An Outline of the General Relativity Theory* TB/510

GEORGE GAMOW: Biography of Physics Σ △ TB/567

STEPHEN TOULMIN & JUNE GOODFIELD: The Fabric of the Heavens: *The Development of Astronomy and Dynamics.* △ *Illus.* TB/579